U0826128

CFforBio “中国森林生物多样性监测网络”丛书 马克平 主编

浙江古田山森林
——树种及其分布格局

Gutianshan Forest Dynamic Plot
Tree Species and Their Distribution Patterns

陈 彬 米湘成 方 腾 陈 磊 任海保 马克平 著

中国林业出版社
China Forestry Publishing House

图书在版编目（CIP）数据

浙江古田山森林——树种及其分布格局／陈彬 等著. —北京：中国林业出版社，2009.9
（“中国森林生物多样性监测网络”丛书）
ISBN 978-7-5038-5667-9
I. 古… II. 陈… III. 森林－树种－研究－开化县 IV. S717.255.4
中国版本图书馆CIP数据核了（2009）第131293号

内容简介

本书介绍了浙江省古田山亚热带常绿阔叶林常见木本植物159种，每个树种除了文字描述外，还配有三张精美的照片，展示植物的小枝、花序、果实和幼苗等，方便识别。同时，附有每种植物在24hm^{2}长期定位研究样地内的种群分布图以及种群的个体数量和径级结构。对于该样地的地形、土壤、植被等皆有介绍。本书以资料翔实、图片精美为特色，是亚热带常绿阔叶林研究不可多得的参考书，也可以作为植物爱好者了解亚热带森林、认识森林植物的野外指导手册。

中国林业出版社·环境景观与园林园艺图书出版中心

策划、责任编辑：于界芬 吴金友
电话：83286967 83229512 传真：83286967

出 版：中国林业出版社（100009 北京西城区德内大街刘海胡同7号）
网 址：www.cfph.com.cn
E-mail：cfphz@public.bta.net.cn 电话：（010）83224477
发 行：新华书店北京发行所
印 刷：北京中科印刷有限公司
版 次：2009年9月第1版
印 次：2009年9月第1次
开 本：889mm × 1194mm 1 / 16
印 张：11.25
字 数：288千字
印 数：1～2000册
定 价：120.00元

序　言1

在过去的几十年时间里，中国科学院和林业、农业等相关部门陆续建立了数百个生态系统定位研究站。其中，中国科学院组建的中国生态系统研究网络(CERN)拥有分布于全国包括农田、森林、草地、湿地、荒漠等生态系统类型的36个生态站。国家林业局建立的中国森林生态系统研究网络(CFERN)由29个生态站组成，基本覆盖了我国典型的地带性森林生态系统类型和最主要的次生林、人工林类型。

随着研究的发展，特别是近年来人们对生物多样性和全球变化研究的关注，国际上正在推动生态系统综合研究网络平台的建立。在全球水平上，全球生物多样性综合观测网络（GEO-BON）是一个有代表性的研究网络。它试图把全球与生态系统和生物多样性长期定位研究相关的网络整合起来，通过综合研究，探讨生态系统与生物多样性维持与变化机制以及系统之间的相互作用机理，为生态系统可持续管理与生物多样性的保护提供科学依据和管理模式。

近年来，中国科学院生物多样性委员会组织建立了中国森林生物多样性监测网络（Chinese Forest Biodiversity Monitoring Network，以下简称 CForBio）。中国是生物多样性特别丰富的少数国家之一，也是唯一一个具有从北部寒温带到南部热带完整气候带谱的国家。CForBio对于揭示中国森林生物多样性形成和维持机制，以及森林生物多样性对全球变化的响应，科学利用和有效保护中国森林生物多样性资源具有重要意义。目前，CForBio已建成4个大于20hm^2的监测样地，分别是长白山阔叶红松林（25 hm^2）、古田山亚热带常绿阔叶林（24 hm^2）、鼎湖山南亚热带常绿阔叶林（20 hm^2）和西双版纳热带雨林（20 hm^2）。广西弄岗、浙江天童、河南伏牛山和北京东灵山的大型监测样地正在建设中。此外，还有若干5 hm^2的样地，如吉林长白山白桦林、北京东灵山辽东栎林、浙江百山祖常绿阔叶林和四川都江堰常绿落叶阔叶混交林样地。CForBio是继美国史密森研究院热带研究所建立的热带森林生物多样性监测网络（CTFS）之后又一大型区域生物多样性监测网络。由于CForBio横跨多个纬度梯度，对揭示生物多样性科学的基本规律有特殊意义，在国际生物多样性监测网络中具有重要地位。

目前，CForBio已经有很好的研究进展，各样地研究成果陆续在国际著名生态学刊物如*Ecology, Journal of Ecology, Oikos*等上发表，受到国内外同行的高度评价。但这些文章都是关于某一具体问题的研究总结，还无法让国内外同行全面了解CForBio各个样地整体情况。因此，出版这套以中英文形式介绍各大样地基本情况的“中国森林生物多样性监测网络”丛书是非常必要的。感谢马克平研究员组织相关专家编写这套丛书。我相信该丛书不仅是国内外同行深入了解CForBio各样地的参考书，同时也将为我国森林生物多样性监测和森林生态系统联网研究奠定重要的基础。

Foreword 1

In the past few decades, hundreds of Ecosystem Research Stations have been set up by the Chinese Academy of Sciences, State Forestry Administration, Ministry of Agriculture and other relative departments. Among them, 36 ecological research stations were established by Chinese Ecosystem Research Network (CERN), supported by the Chinese Academy of Sciences. The 36 research stations are scattered over the country representing diverse ecosystems, including farmland, forest, grassland, wetland, desert and others. Moreover, the Chinese National Ecological Research Network (CFERN), supported by the State Forestry Administration, consists of 29 research stations, covering typical zonal forest ecosystems and main secondary forests and plantations in China.

With the development of research, especially the growing concern over researches on biodiversity and global change in recent years, the establishment of ecosystem research network have been promoted under international supports. So the Group on Earth Observations Biodiversity Observation Network (GEO-BON) is representative across the world, and it attempts to integrate worldwide networks relating to long-term research on ecosystem and biodiversity. Based on the comprehensive studies, the maintenance and change mechanism of ecosystem and biodiversity and their interactions have been explored, which provide scientific basis and management mode for sustainable development of ecosystem and protection of biodiversity.

In recent years, Chinese Forest Biodiversity Monitoring Network (CForBio) has been built by Biodiversity Committee of the Chinese Academy of Sciences. China is one of the few top "mega-biodiversity countries" in the world, and it is also the only country with full climatic zone spectrum, ranging from northern cool temperate zone to southern tropical zone. Besides, CForBio is of great significance to reveal the formation and maintenance mechanism of forest biodiversity in China and their response to climate change

Four permanent plots with area larger than 20-ha have been built at present. They are 25-ha temperate broad-leaved Korean pine mixed forest plot in Changbaishan, 24-ha subtropical evergreen broad-leaved forest plot in Gutianshan, 20-ha lower subtropical evergreen broad-leaved forest plot in Dinghushan and 20-ha tropical rain forest plot in Xishuangbanna. Moreover, other four large-scale permanent forest plots are being built: Longgang in Guangxi, Tiantongshan in Zhejiang, Funiushan in Henan and Donglingshan in Beijing. Besides, a number of 5-ha plots have also been built, including birch forest plot in Changbaishan, Jiling province, oak forest plot in Donglinshan, Beijing, evergreen broad-leaved forest plot in Baishanzu, Zhejiang province and evergreen and deciduous broad-leaved mixed forest plot in Dujiangyan, Sichuan province. Now, CForBio is another regional biodiversity monitoring network after the Center for Tropical Forest Science (CTFS). As being across several latitudinal gradients, CForBio is not only important to examining the fundamental mechanism of biodiversity maintenance, but also plays an important role in Global Biodiversity Monitoring Network.

Encouraging progress has been made in this area since the network built, for lots of research findings have been published in the international peer reviewed ecological journals, such as *Ecology, Journal of Ecology* and *Oikos,* etc., which brought about positive response from colleagues in the field of plant ecology. However, the published papers mostly focus on research of specific problems; scientists and public still can't understand the whole situation of each plot in details. So it is really necessary to publish this series, which introduce basic information of permanent forest plots in both Chinese and English. I am grateful to Professor Keping Ma for organizing related specialists to prepare the series. And I believe that this series would be a valuable reference book for scientists and public to further understand CForBio, and it will also lay a foundation for the forest biodiversity monitoring and forest ecosystem research in China.

Honglie Sun
The former Vice-President for the Chinese Academy of Sciences

序　言2

森林在维持世界气候与水文循环中起着根本性的作用。森林是极为丰富多样的动物、植物与微生物的家园，而人类正是依靠这些生物获取各种产品，包括食品与药物。尽管对人类福祉如此重要，森林仍然遭受着来自土地利用与全球气候环境变化的巨大威胁。在这种不断变化的情况下，为了更好地管理全球剩余的森林，迫切需要树种在生长、死亡与更新方面的详细信息。

中国森林生物多样性监测网络（CForBio）正在中国沿着纬度与环境梯度建立大尺度森林监测样地。通过这个重要的全国行动倡议与来自中国科学院及若干其他单位的研究者的努力，CForBio开始搜集关于中国森林的结构与动态的关键信息。现在CForBio与史密森研究院及哈佛大学阿诺德树木园的热带森林监测网络（CTFS）形成了合作伙伴。CTFS是个在21个热带或温带国家拥有长期大尺度森林动态研究样地的全球性网络。CForBio与CTFS合作的目标是通过合作研究，了解森林是如何运作的，它们是如何随着时间而改变的，以及如何重建或者恢复，以确保森林提供的环境服务能可持续或者增长。森林及其提供的服务的长期可持续性有赖于我们预测森林对全球变化，包括气候与土地利用变化的响应的能力，以及我们去理解与创建适当的森林服务市场的能力。通过拥有34个森林大样地的全球网络及大量项目的训练与能力建设，CForBio与CTFS的伙伴关系是发展这些预测工具的重要基础。这种伙伴关系也将促进为全球各地的当地社区、林业管理者与政策制定者在森林的保育与管理方面发展应用性的林业项目建议，发展与示范利用乡土物种进行森林重建的方法，以及从经济学角度评估森林在减缓气候变化、生物多样性保护和流域保护上的价值的方法。

我祝贺作者们创作了这部关于样地植物的优秀丛书。本丛书为将来的森林监测提供了基准信息，是涉及森林恢复、碳存储、动植物关系、遗传多样性、气候变化、局地与区域保育等研究内容的研究者、学生与森林管理者们有价值的参考资料。

S. J. 戴维斯
主任
史密森热带研究所＆哈佛大学阿诺德树
木园热带森林科学研究中心

Foreword 2

Forests play an essential role in regulating of world's climatic and hydrological cycles. They are home to a vast array of animal, plant and microorganism species on which humans depend for many products, including food and medicines. Despite the importance of forests to human welfare they are under enormous threat from changes in land-use and global climatic conditions. In order to better manage the world's remaining forests under these changing conditions detailed information on the dynamics of growth, mortality and recruitment of tree species is urgently needed.

The Chinese Forest Biodiversity Monitoring Network (CForBio) that aims to establish large-scale forest monitoring plots across latitudinal and environmental gradients in China. Through this important national initiative, researchers from the Chinese Academy of Sciences and several other research institutions in China, CForBio has begun to gather key information on the structure and dynamics of China's forests. The CForBio initiative is now partnering with the Center for Tropical Forest Science (CTFS) of the Smithsonian Research Institute and the Arnold Arboretum of Harvard University. CTFS is a global program of long-term large-scale forest dynamics plots in 21 tropical and temperate countries. The goal of the partnership between CForBio and CTFS is to work together to understand how forests work, how they are changing over time, and how they can be re-created or restored to ensure that the environmental services provided by forests are sustained or increased. The long-term sustainability of forests and the services they provide depend on our ability to predict forest responses to global changes, including changes in climate and land-use, and our ability to understand and create appropriate markets for forest services. The CForBio-CTFS partnership is ideally poised to develop these predictive tools through a global network of 34 large forest plots and an extensive program of training and capacity building. The partnership will also lead to the development of applied forestry programs that advise local communities, forest managers and policy makers around the world on conservation and management of forests, to develop and demonstrate methods of native species reforestation, and to economically value the roles that forests play in climate mitigation, biodiversity conservation, and watershed protection.

I congratulate the authors on the production of this excellent new series of stand books. In addition to providing a baseline for future forest monitoring, these books provide a valuable resource for researchers, students, and forest managers dealing with issues of forest restoration, carbon storage, plant-animal interactions, genetic diversity, climate change, and local and regional conservation issues.

Stuart Davies
Director
Center for Tropical Forest Science
The Smithsonian Tropical Research Institute &
The Arnold Arboretum of Harvard University

前　言

我国的常绿阔叶林在世界范围内分布面积最广，发育最为典型，为同类植被中罕见的类型，其中亚热带是常绿阔叶林的典型分布区，其分布范围约位于北纬24°～32°，东经99°～123°之间，是我国重要植被类型之一。

我国的研究工作者在亚热带常绿阔叶林区逐步建立了各具区域特色的常绿阔叶林研究基地，对常绿阔叶林的物种组成和空间分布结构有一定的研究积累。但这些研究的样地面积小、监测时间短，难于涵盖众多稀有物种，较难反映群落内不同尺度上的生物多样性格局和过程，不足以认识群落水平上生物多样性维持机制，因此有必要在亚热带典型常绿阔叶林地带建立大面积长期监测样地，监测常绿阔叶林从种子雨散布、幼苗更新和成株的变化动态过程，以及动物和微生物类群的变化及其功能，探索常绿阔叶林的生物多样性形成和维持机制。

浙江省古田山国家级自然保护区具有亚热带常绿阔叶林丰富的种类和独特的类群，目前还保存着很好的低海拔天然植被，这在中亚热带东部地区十分少见，而且古田山自然保护区也是国际生物多样性保护的热点地区，在该自然保护区选择典型的地带性植被甜槠-木荷林作为长期监测样地，监测和研究常绿阔叶林的生物多样性形成和维持机制，具有典型性和代表性。基于上述原因，生物多样性委员会组织中国科学院植物研究所、浙江大学、古田山国家级自然保护区管理局等单位，采用CTFS（Centre for Tropical Forest Science）样地建设标准，于2004～2005年建成了24hm^2的永久监测样地，以期对古田山常绿阔叶林生物多样性进行长期监测与研究。我国是世界上唯一跨越多个气候带的国家，古田山样地和长白山样地、鼎湖山样地、西双版纳样地等分别代表不同气候带的地带性植被，同时成为中国森林生物多样性长期监测网络的核心组成部分。

本书详尽地描述了古田山24 hm^2样地常绿阔叶林群落的物种特性、分布格局和径级结构，而精美的样地植物照片让读者对本书的内容有更感性的认知。同时本书也是古田山样地建设启动以来5年左右工作积累的展现，为未来样地的深入研究提供了必备的基础生态信息，也期待吸引更多的年轻学子加入森林生态学的探索行列。

在地形复杂的古田山建设大型的长期监测样地，是一个前所未有的挑战。在样地建设的过程中得到了方方面面的大力帮助。借此机会，诚挚地感谢下列单位和同事对样地建设所作出的贡献：

(1) 提供科学及技术支持的加拿大阿尔伯塔大学何芳良教授、东海大学孙义方教授、温州大学丁炳扬教授、浙江大学于明坚副教授和丁平教授、浙江师范大学陈建华副教授、哈佛大学Stuart Davies 博士；

(2) 提供工作和生活便利的古田山国家级自然保护区管理局胡青延局长、钱海源副局长、郑球星副局长、吕一锋副局长、余建平主任、郑东红主任、方藤工程师、陈声文工程师、尤登法、詹翠玉、程凌宏、吴应红和开化县苏庄镇姜瑞根；

(3) 提供精美照片的陈彬、方藤、金冬梅、张金龙、陈国科、陈建华、杨波、贾琪、宋凯、刘晓娟、邴艳红、郑小军、何祖霞、喻勋林、杜彦君、李晓东、严岳鸿、易思荣、张代贵等；

(4) 参加样地建设的中国科学院植物研究所米湘成、任海保、官贵权、黄继红、张俪文，浙江大学于明坚、赵谷风、沈国春、慎佳泓、张洋、张磊、黄军、余佳、王琰琰、苏莹雪、胡轭，浙江师范大学陈建华、黄敏、陈庆杰、夏振国、丁高兴、黄秋生、李立、李侠、宋瑞生、杨武、蒋华伟，中国计量学院唐乔、李力霖，温州师范学院丁炳扬、姜雪华，开化县苏庄镇赖祯熙、姜智勇。

时间仓促，水平所限，错误疏漏在所难免，敬请各位读者不吝赐教。

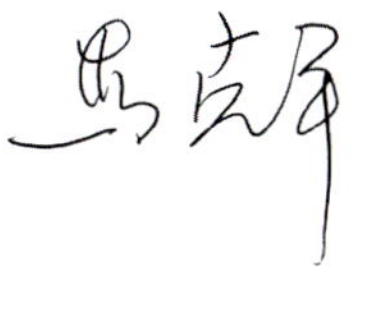

Preface

China is the central area of evergreen broad-leaved forest in the world, and evergreen broad-leaved forest in China is characterized by its typical development and vast distributional area in southwestern and southeastern part of China. As one of important vegetation types in China, evergreen broad-leaved forest is mainly distributed in subtropical zone ranging from 24° -32° N to 99° -123° E.

Based on observational stations in different parts of evergreen broadleaved forests, Chinese ecologists have studied evergreen broadleaved forests for decades and had many publications in species composition, spatial distribution and so on. However, these studies were usually biased to short period observations on small plots (smaller than 1 ha) which are unable to includes the large number of rare species in this region. Therefore, the studies can hardly reflect patterns and processes of evergreen broadleaved forests across scales, and are not sufficient to shed light on mechanisms of biodiversity maintenance at community level. In order to explore mechanisms of biodiversity origin and maintenance, it is imperative to establish long-term large scale forest dynamic plot and to monitor the dynamic processes from seed dispersal and seedling recruitment to adult, and interactions among forests, animals and microbes.

As an international biodiversity hot spot, Gutianshan National Nature Reserve processes well-protected old-growth evergreen broadleaved forests at low altitude, which can hardly be found in eastern area of subtropical zone. These forests are rich in representative species and community types of evergreen broadleaved forests. It is typical and representative to study mechanisms of biodiversity origin and maintenance based on forest dynamic plot (FDP) in Castanopsis eyrei-Schima superba forest of the Reserve. By these lines of considerations, Biodiversity Committee of Chinese Academy of Sciences organized Institute of Botany, Zhejiang University and Administrative Bureau of Gutianshan Nature Reserve to establish Gutianshan 24 ha FDP following the standard census protocol of CTFS (Center for Tropical Forest Science). China is unique country across multiple climate zones in the world, and Gutianshan 24-ha FDP will be core component together with Changbaishan 25-ha FDP, Dinghushan 20-ha FDP and Xishuangbanna 20-ha FDP representing zonal vegetations across climate zones and constituting the Chinese Forest Biodiversity Monitoring Networks (CForBio).

We have attempted to describe the species characteristics, distribution pattern and DBH size classes found in Gutianshan FDP in this book. This book contains a large number of beautiful photos of seedlings, flowers, fruits and seeds and adult trees accumulated in the five-year studies during and after the plot establishment. This book provides baseline eco-information for future studies, and attempt to attract more youths to join the exploration of forest ecology.

It is an unprecedented challenge to establish large scale forest dynamic plot in terrain-rugged Gutianshan National Nature Reserve. Were it not for their joint contributions and persistence, Gutianshan FDP could not have been established. Hence, the authors greatly acknowledge the following people and institutions for their admirable contributions to the establishment of Gutianshan FDP:

(1) Professor Fangliang He at University of Alberta, Canada, Professor I-Fang Sun at Tunghai University, Professor Ding Binyang at Wenzhou University, Associate Professor Yu Mingjian and Professor Ding Ping at Zhejiang University, Associate Professor Chen Jianhua at Zhejian Normal University, Dr. Stuart Davies at Harvard University for their scientific and technique support;

(2) Director General Hu Qingyan, Deputy Director General Qian Haiyuan, Deputy Director General Zheng Qiuxing, Deputy Director General Lu Yifeng, Director Yu Jianping, Director Zheng Donghong, Egnineer Fang Teng, Egineer Cheng Shengwen, You Dengfa, Zhang Cuiyu, Cheng Linhong, Wu Yinghong at Administrative Bureau of Gutianshan National Nature Reserve, and Jiang Ruigeng at Suzhuang Town, Kaihua County for adiministrative and logisitic support;

(3) Chen Bin, Fang Teng, Jin Dongmei, Zhang Jinlong, Chen Guoke, Chen Jianhua, Yang Bo, Jia Qi, Liu Xiaojuan, Bing Yanhong, Zheng Xiaojun, He Zuxia, Yu Xunlin, Du Yanjun, Li Xiaodong, Yan Yuehong, Yi Sirong, Zhang Daigui for providing beautiful photos;

(4) Mi Xiangcheng, Ren Haibao, Gong Guiquan, Huang Jihong, Zhang Liwen at Institute of Botany, Chinese Academy of Sciences, and Yu Mingjian, Zhao Gufeng, Sheng Guochun, Shen Jiahong, Zhang Yang, Zhang Lei, Huang Jun, Yu Jia, Wang Yanyan, Su Yinxue, Hu E at Zhejiang University, Chen Jianhua, Huang Min, Chen Qingjie, Xia Zhengguo, Ding Gaoxin, Huang Qiusheng, Li Li, Li Xia, Song Ruisheng, Yang Wu, Jiang Weihua at Zhejiang Normal University, Tang Qiao Li Lilin at Chinese Jiliang University, Jiang Xuehua at Wenzhou Normal College, and Lai Zhenxi and Jiang Zhiyong at Suzhuang Town, Kaihua County for their participation of plot establishment.

Ma Keping

目　录

Contents

古田山24hm²样地简介

1. 古田山自然保护区

1.1 地理位置和自然环境

古田山国家级自然保护区（29°10′19.4″～29°17′41.4″N，118°03′49.7″～118°11′12.2″E），位于浙江省开化县西部的苏庄镇境内，总面积8107hm²，地处亚热带季风气候区，同时由于山势高峻，山体东北-西南走向，使湿润的夏季风能长驱直入，带来丰沛的降水，而干燥寒冷的冬季风则被山体阻挡，从而形成冬暖夏凉、温暖湿润、雨量充沛的气候特点。年平均温度15.3℃，年平均降水1963.7mm，每年的3～6月为第一雨季，7～8月为伏旱季，8月底至9月受台风影响形成第二雨季，10～12月为干期。

古田山土壤基本上是由母岩风化形成，主要有红壤、黄红壤、红黄壤和沼泽土4大类，其中红壤、黄红壤、红黄壤占全区总面积80%以上，分别分布于海拔500～700m，700～1000m和1000m以上。土壤土表残落物层较薄，C/N比不高，绝大多数土壤呈酸性反应。

Introduction to Gutianshan 24-ha Forest Dynamic Plot

1. Introduction to Gutianshan Nature Reserve

1.1 Location and description of Gutianshan Nature Reserve

Gutianshan National Nature Reserve (29°10′19.4″~29°17′41.4″N, 118°03′49.7″~118°11′12.2″E), approximately 8,107 hectares, is located in Kaihua County, Zhejiang Province in east China. The local climatic characteristic is formed by interaction of subtropical monsoon climate and the land form, mainly the high and steep mountains. In summer, warm and mild wind from northeast brings rich precipitation; whereas in winter, cold and dry wind is blocked out and has limited influence on local climate. Annual mean temperature in the region is 15.3

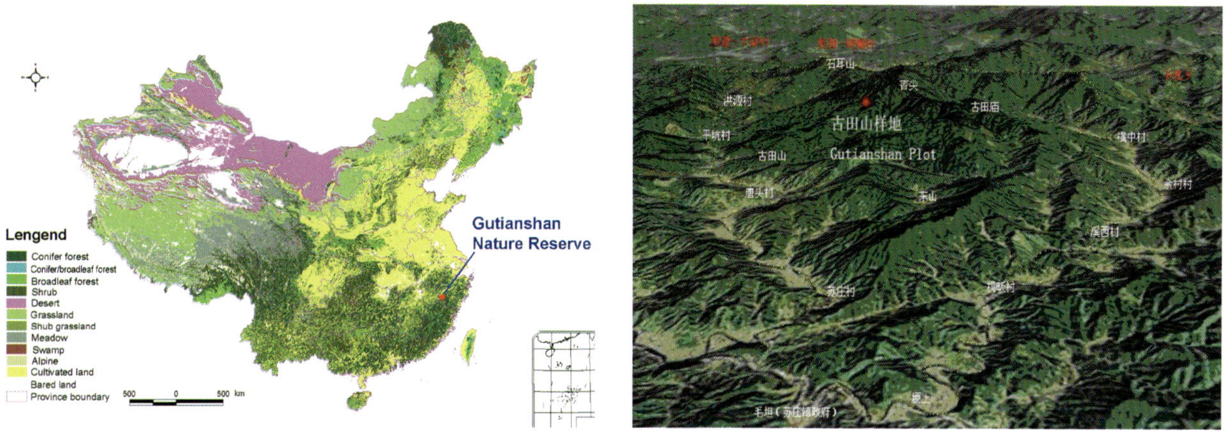

A　　　　B

古田山自然保护区的位置及24hm²样地在古田山保护区中的位置

A. 古田山自然保护区的位置；B. 24hm²样地在古田山保护区中的位置

Map of China showing the location of the 24-ha Gutianshan permanent plot, eastern China.

A. The location of Gutianshan Nature Reserve in China; B. The location of 24 ha forest plot in Gutianshan Nature Reserve.

centidegree, annual mean precipitation, calculated from data of years from 1958 to 1986, is 1963.7mm, majority of which occurs between March and September.

Weathering and denudation of the mother rock ultimately formed 4 central types of soil in Gutianshan: red soil, yellow soil, red-yellow soil and yellow-red soil, the former three account for more than 80% of the total area, and are mainly distributed in 500~700m, 700~1000m, above 1000m a.s.l., respectively. The litter cover is generally thin, the carbon-nitrogen ratio is usually narrow, and the vast majority of soil shows acid reaction.

1.2 主要植被类型

常绿阔叶林是古田山分布面积最广的植被类型，主要分布于海拔350～800m的山坡和山麓；在海拔720m（山脊）和800m（山谷）至海拔1100m左右，大片分布着黄山松针阔叶混交林，海拔1100m以上至山顶为黄山松林。主要类型包括：

I. 常绿阔叶林（海拔400～900m）

（1）甜槠林（主要类型）：a. 甜槠－青冈林；b. 甜槠－石栎林；c. 甜槠－木荷林；

（2）栲树林；

（3）野含笑－钩栗林；

（4）青冈林；

（5）虎皮楠－甜槠林；

（6）乌冈栎－青冈林。

II. 常绿落叶阔叶混交林

（1）山地常绿落叶阔叶混交林：短柄枹－细叶青冈林；

（2）沟谷常绿落叶阔叶混交林：a. 南酸枣－紫楠林；b. 红毒茴－紫茎林。

III. 暖性针阔叶混交林

（1）马尾松－木荷－甜槠林；

（2）马尾松－青冈林。

IV. 温性针阔叶混交林

（1）黄山松－木荷林；

（2）黄山松－细叶青冈林。

V. 暖性针叶林

马尾松林。

VI. 温性针叶林

黄山松林。

1.2 Main Vegetation Types

Most of the region is covered by ever-green broadleaved forest, mostly distributed at altitude of 350 to 800 m on hillside and foothill. Vegetation on the ridge from 720 to about 1100 m and in the valley between 800 and 1100 m is mainly the deciduous broadleaved and coniferous mixed forest, and 1100 m above is *Pinus taiwanensis* forest. The following are main vegetation types:

I. Evergreen broadleaved forest (Altitude 400-900m)

(1) *Castanopsis eyrei* Forest (Main type): a. *Castanopsis eyrei-Cyclobalanopsis glauca* forest; b. *Castanopsis eyrei-Lithocarpus glaber* forest; c. *Castanopsis eyrei-Schima superba* forest;

(2) *Castanopsis fargesii* Forest;

(3) *Manglietia skinnneriana-Castanopsis tibetan* forest;

(4) *Cyclobalanopsis glauca* forest;

A B C D

古田山自然保护区不同的森林类型
Different types of forests in Gutianshan Nature Reserve
A. 常绿阔叶林 Evergreen broadleaved forest; B.常绿落叶阔叶混交林 Evergreen and deciduous broadleaved mixed forest; C. 针阔混交林 Coniferous and broadleaved mixed forest; D.针叶林 Coniferous forest

(5) *Daphniphyllum oldhami-Castanopsis eyrei* forest;

(6) *Quercus phillyraeoides-Cyclobalanopsis glauca* forest.

II. Evergreen broadleaved and deciduous broadleaved mixed forest

(1). Mountain evergreen broadleaved and deciduous broadleaved mixed forest;
Quercus serrata-Cyclobalanopsis gracilis forest:

(2). Ravine evergreen broadleaved and deciduous boradleaved mixed forest: a. *Choerospondias axillaris-Phoebe sheareri* forest; b. *Illicium lanceolatum- Stewartia sinensis* forest.

III. Warm temperate deciduous broadleaved and coniferous mixed forest

(1) *Pinus massoniana-Schima superba-Castanopsis* eyrei forest;

(2) *Pinus massoniana-Cyclobalanopsis glauca* forest.

IV. Temperate deciduous broadleaved and coniferous mixed forest

(1) *Pinus taiwanensis-Schima superba* forest;

(2) *Pinus taiwanensis-Cyclobalanopsis gracilis* forest.

V. Warm temperate coniferous forest

Pinus massoniana forest.

VI. Temperate coniferous forest

Pinus taiwanensis forest.

2. 古田山24hm^{2}森林样地

2.1 样地建设与群落调查

样地植被为典型的中亚热带低海拔常绿阔叶林，位于浙江省古田山国家级自然保护区内，南北宽400m、东西长600m，总面积24 hm^2。2004年11月开始样地建设，2005年9月完成第一次群落调查。样地建设和群落调查参照CTFS（Centre for Tropical Forest Sciences）的方法（http://www.ctfs.si.edu/），用全站仪将样地划分为600个20m × 20m的样方，标定并调查样方内所有胸径（DBH）≥1 cm的木本植物个体，内容包括植物个体的物种名称、胸径、位置等，并挂牌标记。

2. Gutianshan 24-ha forest plot

2.1 Plot Establishment and Plant Census

The forests in Gutianshan plot is representative of mid-subtropical evergreen broadleaved forest in Gutianshan National Nature Reserve. Gutianshan plot ranges 600 m from east to west, and 400 m from south to north. According to CTFS's plot census method (http://www.ctfs.si.edu/), the total plot had been grided into 600 20 m × 20 m subplots. All trees with DBH ≥ 1 cm were tagged, identified, measured, and georeferenced, corresponding data such as species, DBH, location data of each tagged tree was recorded.

2.2 地形和土壤

古田山样地地形复杂多变，最高海拔714.9m，最低海拔446.3m，平均海拔580.6m，最大高差268.6m。样地地势较陡，坡度范围12.8°～62°，平均坡度约37.5°。土壤类型主要为红壤，pH值在5.5～6.5之间。

2.2 Topography and Soil

The plot is characterized by rugged terrain: altitude varies from 446.3m to 714.9m above sea level, and the mean elevation is 580.6 m. The mean slope in plot is 37.5° and 20-m cell slope ranges from 12.8° to 62°. Soil type of the plot consists mainly of red soil (equivalent to Ultisols in US soil taxonomy), and the pH value is between 5.5 and 6.5.

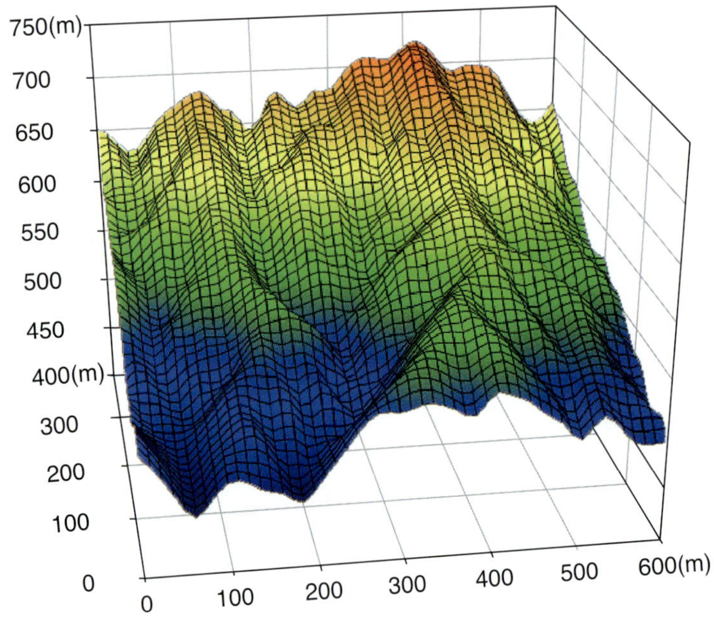

古田山样地三维地形图

Topographical map of Gutianshan plot

样地为长方形，南北宽400m（y轴），东西长600m（x轴）；图中数字为相对高程，以样地最低点为 0 m。 The plot is a rectangle, horizontal axis is 400m(SN), vertical axis is 600m(EW); the number in the figure is relative height, 0 m is the lowest point.

2.3 物种组成与群落结构

样地共有140700个胸径（DBH）≥1 cm的独立的木本植物个体，分属49科，104属，159个种。泛热带科比例最大，为28.6%，代表科为樟科、山茶科和卫矛科；其次为北温带分布科，占20.4%，代表科为壳斗科、山茱萸科。属的类型仍以泛热带分布最多，占到21.8%。种的分布特点与科的分布特点相似。样地的区系组成热带分布类型多于温带分布类型。

样地优势物种明显，以甜槠(*Castanopsis eyrei*)、木荷(*Schima superba*)为主要优势物种，32个物种个体数大于1000；样地内稀有种丰富，59个物种平均每公顷个体数<1，占总物种数的37.1%。

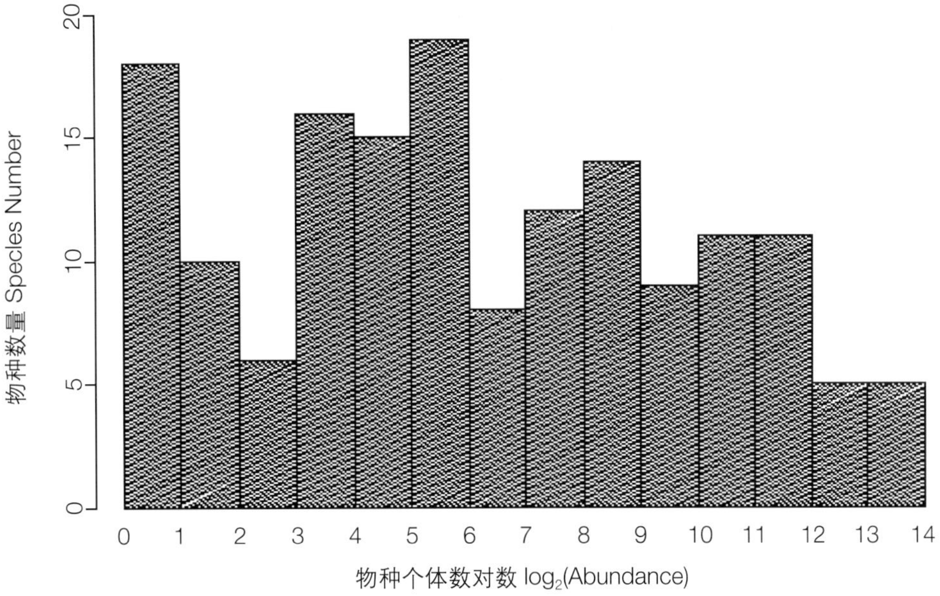

古田山24hm²样地物种－个体数分布图（物种个体数取以2为底的对数）
Species-abundance distribution of Gutianshan 24ha plot (The abundance is log2 transformed)

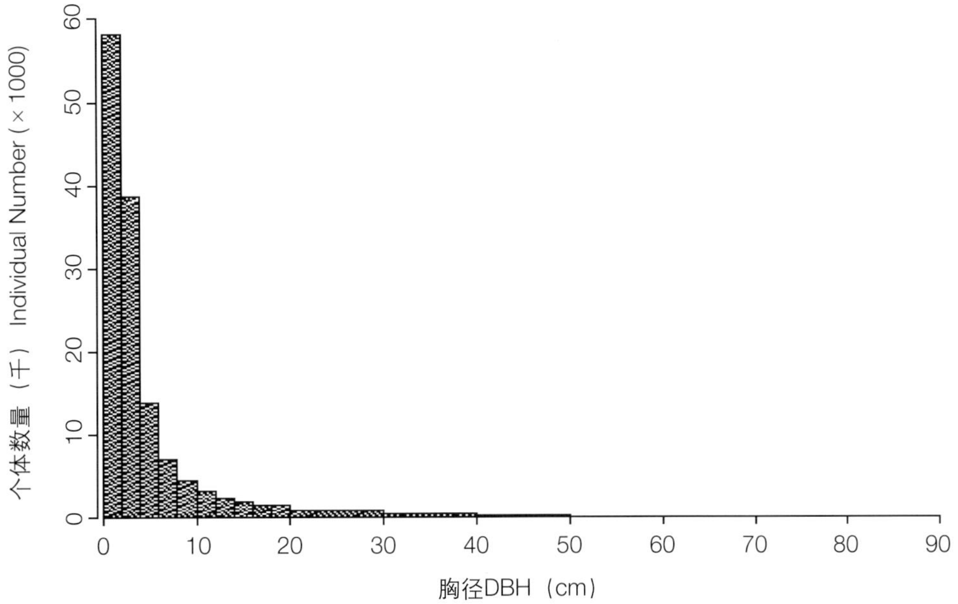

古田山24hm² 样地胸径≥1cm木本植物个体径级分布图
DBH distribution of individuals with DBH≥1cm in Gutianshan 24 ha plot

样地群落垂直结构清晰，分为乔木层、亚乔木层、灌木层。乔木层以甜槠、木荷、马尾松(*Pinus massoniana*)为优势物种，亚乔木层以枹栎（短柄枹）(*Quercus serrata*)、马银花(*Rhododendron ovatum*)、浙江新木姜子(*Neolitsea aurata* var. *chekiangensis*)为优势物种，灌木层以柳叶蜡梅(*Chimonanthus salicifolius*)、映山红(*Rhododendron simsii*)等为优势物种。

样地内木本植物个体最大胸径为87.4 cm，平均胸径为5.2 cm。样地木本植物个体径级总体呈倒“J”型，小径级个体数量较大，胸径1～4 cm的个体数量占总个体数量的69.07%。

2.3 Species Composition and Community Structure

140,700 individuals (DBH $\geqslant$ 1 cm) were recorded in the plot, belongging to 49 families, 104 genera and 159 species. At family level, the proportion of the Pantropic type is the greatest (28.6%), of which Lauraceae, Theaceae, Celastraceae are typical. The second is North Temperate type (20.4%), of which Fagaceae and Cornaceae are main types. At genus level, Pantropic type is also the greatest (21.8%). The characteristics of flora at species level are similar to that at family level. Overall, there are more floristic elements in the plot consisting tropical type than temperate one.

In the plot, the forest communities are dominated by *Castanopsis eyrei* and *Schima superba*, and 32 species have more than 1000 individuals. *In the plot*, the rare species, i.e. less than 1 individual per ha, account for 37.1% (59 rare species/total species) of total species richness.

The vertical structure of the forest is clear: canopy story, sub-canopy story and shrub story. Canopy story is dominated by *Castanopsis eyrei*, *Schima superba* and *Pinus massoniana*; sub-canopy story by *Quercus serrata*, *Rhododendron ovatum* and *Neolitsea aurata* var. *chekiangensis*; and shrub story by *Chimonanthus salicifolius* and *Rhododendron simsii*.

The maximal DBH is 87.4 cm and the mean value of BDH is 5.2 cm in the plot. The DBH size of individuals in the plot follows the distribution of inverse “J” shape. Small individuals account for the most of abundance. For example, the individuals with DBH from 1 to 4 cm accounts for 69.07% of the total abundance.

1 马尾松 mǎwěisōng | Masson Pine

Pinus massoniana Lamb.
松科 | Pinaceae

代码（SpCode）= PINMAS
个体数（Individual number/24ha）= 2061
最大胸径（Max DBH）= 78.2 cm
重要值排序（Importance value rank）= 23

常绿乔木，高达45m。喜光，是干扰后裸地植被演替的先锋树种。针叶2针一束，罕3针1束。树干皮红褐色至灰褐色，呈不规则开裂并片状剥落。球果幼时绿色，成熟时褐色，卵形。子叶5～8枚。花期4月，球果翌年10～12月成熟。

Evergreen conifer, up to 45 m tall; heliophilous, a pioneer species of the succession from denuded land after disturbance. Needles 2(or seldomly 3) per bundle. Bark gray or red-brown, irregularly scaly and flaking. Seed cones green when young, turning chestnut brown at maturity, ovoid in shape. Cytyledons 5-8. Pollination Apr.-May, seed maturity Oct.-Dec. of the following year.

树干 Trunk
摄影：方腾 Photo by: Fang Teng

雄球花 Male cones
摄影：何祖霞 Photo by: He Zuxia

幼球果 Cones(immature)
摄影：何祖霞 Photo by: He Zuxia

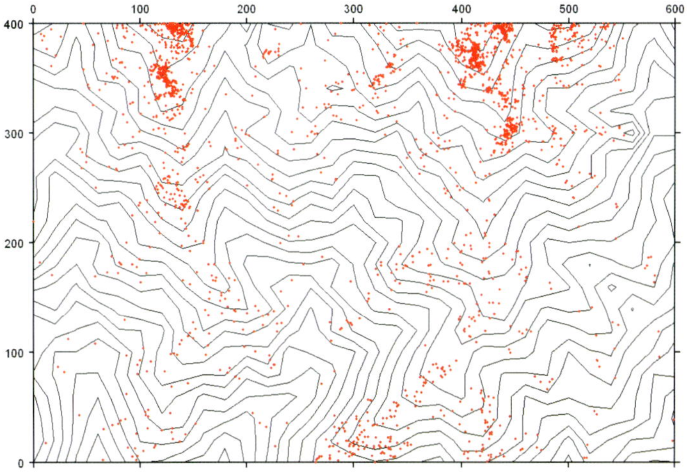

个体分布图 Distribution of Individuals

径级分布表 DBH Class

胸径区间 (Diameter class) (cm)	个体数 (No. of individuals in the plot)	比例 (Proportion)
1～2	268	0.13
2～5	379	0.18
5～10	261	0.13
10～20	332	0.16
20～30	250	0.12
30～60	550	0.27
≥60	21	0.01

2 杉木 shāmù | China Fir

Cunninghamia lanceolata (Lamb.) Hook.
杉科 | Taxodiaceae

代码（SpCode）= CUNLAN
个体数（Individual number/24ha）= 9
最大胸径（Max DBH）= 39 cm
重要值排序（Importance value rank）= 133

常绿乔木，高达30m。主干通直，单轴分枝。侧枝轮生。叶先端尖而硬。雄球花簇生枝顶；雌球花单生，或2～3朵簇生枝顶，卵圆形。球果近球形或圆卵形；子叶2枚。花期4月，果期10～11月。非本地野生树种。

Evergreen conifer, up to 30 m tall; trunk straight, monopodial and whorled branching. Leaves apex usually spinescent. Pollen cone fascicles terminal, seed cones solitary or 2-3 together. Cytyledons 2. Pollination Apr., seed maturity Oct.-Nov.. Not local wild species.

植株 Whole plant
摄影：陈彬 Photo by: Chen Bin

雄球花 Male cones
摄影：陈彬 Photo by: Chen Bin

雌球花 Female strobili
摄影：陈彬 Photo by: Chen Bin

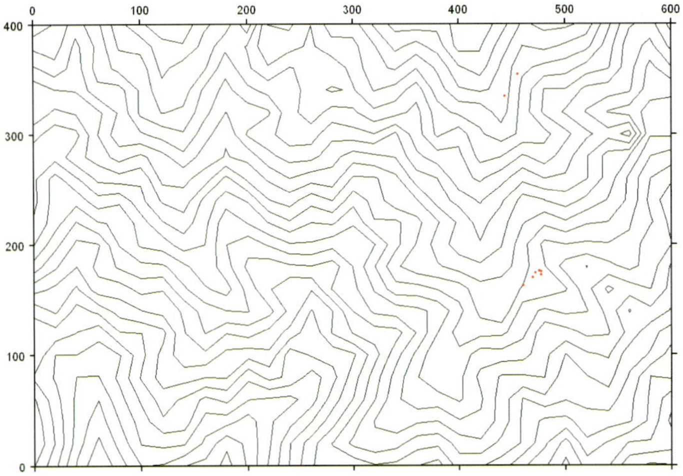

个体分布图 Distribution of Individuals

径级分布表 DBH Class

胸径区间 (Diameter class) (cm)	个体数 (No. of individuals in the plot)	比例 (Proportion)
1～2	1	0.11
2～5	1	0.11
5～10	1	0.11
10～20	3	0.33
20～30	0	0.00
30～60	3	0.33
≥60	0	0.00

3 杨梅 yángméi | China Waxmyrtle; China Bayberry

Myrica rubra (Lour.) Sieb. et Zucc.
杨梅科 | Myricaceae

代码（SpCode）= MYRRUB
个体数（Individual number/24ha）= 907
最大胸径（Max DBH）= 66 cm
重要值排序（Importance value rank）= 24

常绿乔木，高达15m。树皮灰色。树冠圆球形。叶革质，常密生小枝上部。花雌雄异株。核果球形，成熟时深红色到紫红色，外表具长达3mm的乳头状突起，可食。花期3～4月，果期5～7月。

Evergreen tree, up to 15 m tall; bark gray. Canopy spherical. Leaves leathery, acumulate on top of twigs. Flowers dioecious. Drupe globose, dark red or purple-red at maturity, papillae to 3 mm, edible. Fl. Mar.-Apr., fr. May-Jul..

幼苗 Seedling
摄影：方腾 Photo by: Fang Teng

雄花序 Male inflorescence
摄影：方腾 Photo by: Fang Teng

核果 Drupe
摄影：方腾 Photo by: Fang Teng

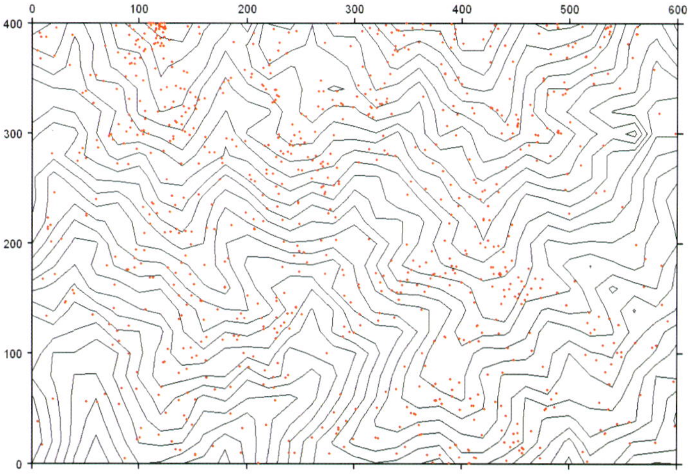

个体分布图 Distribution of Individuals

径级分布表 DBH Class

胸径区间 (Diameter class) (cm)	个体数 (No. of individuals in the plot)	比例 (Proportion)
1～2	173	0.19
2～5	199	0.22
5～10	172	0.19
10～20	220	0.24
20～30	109	0.12
30～60	33	0.04
≥60	1	0.00

4 化香树 huàxiāngshù | Dyetree

Platycarya strobilacea Sieb. et Zucc.
胡桃科 | Juglandaceae

代码（SpCode）= PLASTR
个体数（Individual number/24ha）= 89
最大胸径（Max DBH）= 20.5 cm
重要值排序（Importance value rank）= 73

落叶灌木到乔木，高达15m；树皮灰色，老时不规则纵裂。奇数羽状复叶，小叶对生，具7～23枚小叶。两性花序和雄花序生于小枝顶端。果序球果状，常宿存。花期5～6月，果期7～11月。

Deciduous shrub or tree, up to 15 m tall. Leaf odd-pinnate, leaflets opposite, 7-23. Androgynous spike and male spikes at twig apex. Nutlet cone suborbicular to obovate, usually persistent. Fl. May-Jun., fr. Jul.-Nov..

幼苗 Seedling
摄影：金冬梅 Photo by: Jin Dongmei

花枝 Flowering branch
摄影：陈彬 Photo by: Chen Bin

果序 Infructescence
摄影：方腾 Photo by: Fang Teng

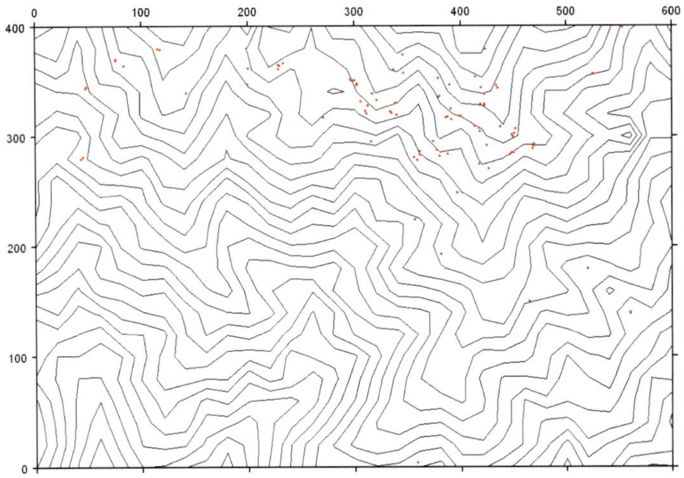

个体分布图 Distribution of Individuals

径级分布表 DBH Class

胸径区间 (Diameter class) (cm)	个体数 (No. of individuals in the plot)	比例 (Proportion)
1～2	2	0.02
2～5	34	0.38
5～10	40	0.45
10～20	12	0.13
20～30	1	0.01
30～60	0	0.00
≥60	0	0.00

5 雷公鹅耳枥 léigōngéěrlì | Thunder God Hornbeam

Carpinus viminea Lindl.
桦木科 | Betulaceae

代码（SpCode）= CARVIM
个体数（Individual number/24ha）= 6
最大胸径（Max DBH）= 25 cm
重要值排序（Importance value rank）= 127

落叶乔木，高10～20m，树皮深灰色；小枝棕褐色，密生白色皮孔，无毛。叶柄细长。叶片侧脉12～15对，近无毛。果序长5～15cm，下垂。花期3～4月，果期5～9月。

Deciduous tree 10-20 m tall; bark dark gray. Branchlets brown, glabrous, with white and dense lenticels. Petiole slender; leaf blade almost glabrous, lateral veins 12-15; fruiting clusters 5-15 cm, drooping. Fl. Mar.-Apr., fr. May-Sep..

小枝 Branchlet
摄影：陈建华 Photo by: Chen Jianhua

雄花序 Male inflorescence
摄影：陈建华 Photo by: Chen Jianhua

果序 Infructescence
摄影：方腾 Photo by: Fang Teng

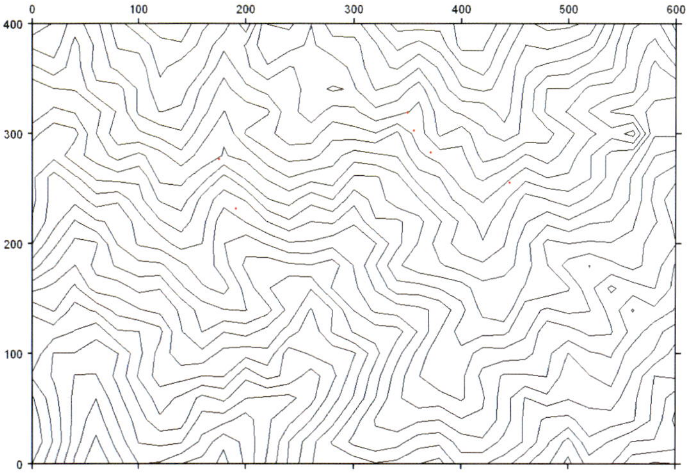

个体分布图 Distribution of Individuals

径级分布表 DBH Class

胸径区间 (Diameter class) (cm)	个体数 (No. of individuals in the plot)	比例 (Proportion)
1～2	1	0.17
2～5	0	0.00
5～10	0	0.00
10～20	4	0.67
20～30	1	0.17
30～60	0	0.00
≥60	0	0.00

6 米槠

mǐzhū | Carles's Oatchesnut

Castanopsis carlesii (Hemsl.) Hay.

壳斗科 | Fagaceae

代码（SpCode）= CASCAR

个体数（Individual number/24ha）= 83

最大胸径（Max DBH）= 41.6 cm

重要值排序（Importance value rank）= 68

常绿乔木，高达20m。小枝黑褐色，皮孔多。叶全缘或有少数浅齿，叶背银灰色或多少带灰白色。圆锥花序近顶生。壳斗近圆球形或阔卵形，长10～15mm，外壁有突起或短刺。花期3～6月，果翌年9～11月成熟。

Evergreen tree, up to 20 m tall; branchlets dark brown, with many lenticels. Leaf margin entire or with a few shallow teeth, back silver grey or hoar. Panicles nearly terminal; cupule subglobose to ovoid, 1-1.5 cm, outside spinelike or reduced to tubercles. Fl. Mar.-Jun., fr. Sep.-Nov. of the following year.

幼苗 Seedling

摄影：金冬梅 Photo by: Jin Dongmei

小枝 Branchlet

摄影：方腾 Photo by: Fang Teng

果序 Infructescence

摄影：喻勋林 Photo by: Yu Xunlin

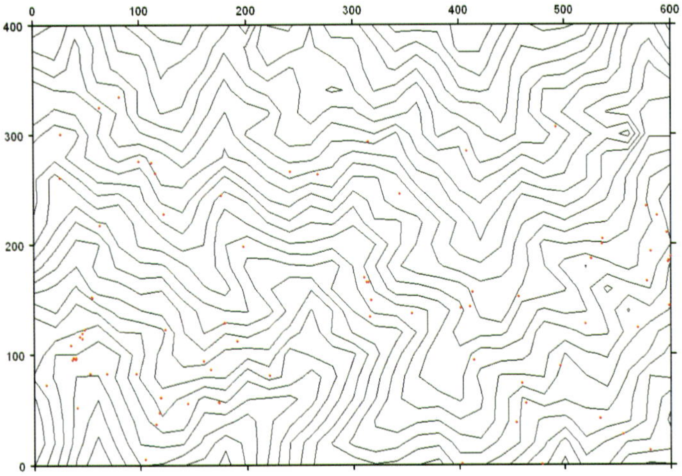

个体分布图 Distribution of Individuals

径级分布表 DBH Class

胸径区间 (Diameter class) (cm)	个体数 (No. of individuals in the plot)	比例 (Proportion)
1～2	28	0.34
2～5	29	0.35
5～10	15	0.18
10～20	5	0.06
20～30	4	0.05
30～60	2	0.02
≥60	0	0.00

7 甜槠 tiánzhū | Sweet Oachestnut

Castanopsis eyrei (Champ. ex Benth.) Tutcher
壳斗科 | Fagaceae

代码（SpCode）= CASEYR
个体数（Individual number/24ha）= 12405
最大胸径（Max DBH）= 87.4 cm
重要值排序（Importance value rank）= 1

常绿乔木，高达20m。树皮纵深裂。枝叶无毛。当年生叶两面同色，二年生叶背面常带淡薄的银灰色。壳斗连刺直径2～3cm，刺长4～10mm。花期4～6月，果翌年9～11月成熟。

Evergreen tree, up to 20 m tall. Bark has deep longitudinal cracks; branches and leaves glabrous. The 1st year leaves with the same color of both sides and the the 2ed year ones are lightly silver grey on back. Cupule 2-3 cm in diam, bracts spinelike, 4-10 mm. Fl. Apr.-Jun., fr. Sep.-Nov. of the following year.

树皮 Bark
摄影：方腾 Photo by: Fang Teng

花序 Inflorescence
摄影：金冬梅 Photo by: Jin Dongmei

果序 Infructescence
摄影：张金龙 Photo by: Zhang Jinlong

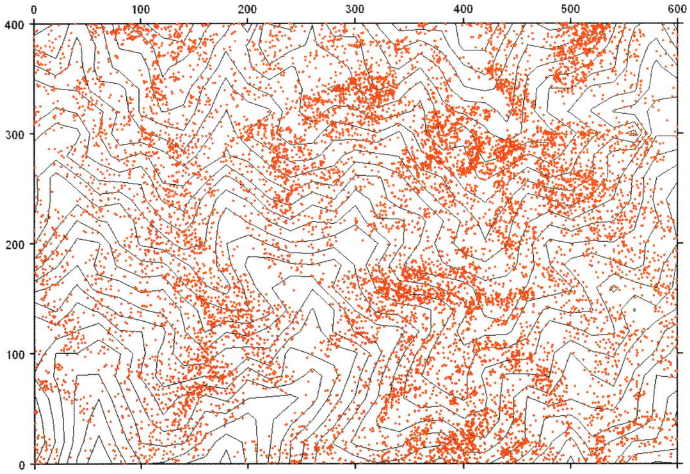

个体分布图 Distribution of Individuals

径级分布表 DBH Class

胸径区间 (Diameter class) (cm)	个体数 (No. of individuals in the plot)	比例 (Proportion)
1～2	3945	0.32
2～5	2127	0.17
5～10	1348	0.11
10～20	1858	0.15
20～30	1649	0.14
30～60	1425	0.11
≥60	24	0.00

8 栲树（栲、丝栗栲） kǎoshù(kǎo、sīlìkǎo) | Farges's Oatchestnut

Castanopsis fargesii Franch.
壳斗科 | Fagaceae

代码（SpCode）= CASFAR
个体数（Individual number/24ha）= 1231
最大胸径（Max DBH）= 50.5 cm
重要值排序（Importance value rank）= 29

常绿乔木，高10～30m。树皮浅纵裂。嫩叶叶背红褐色、黄褐色或灰白色，成长叶叶背黄棕色。雌花序穗状。壳斗通常圆球形或宽卵形，连刺直径2.5～3cm，刺长8～10mm。花期4～6月、8～9月，果期翌年4～9月。

Evergreen tree 10-30 m tall; Bark has shallow longitudinal cracks. Leaf blade oblong reddish brown to yellowish brown or greyish-white when young, but tawny with age. Female inflorescences solitary; cupules scattered on rachis, globose to broadly ovoid, 2.5-3 cm in diam, bracts spinelike, 8-10 mm. Fl. Apr.-Jun. and Aug.-Sep., fr. Apr.-Sep. of the following year.

叶背 Leaf back
摄影：陈建华 Photo by: Chen Jianhua

花序 Inflorescence
摄影：陈彬 Photo by: Chen Bin

果序 Infructescence
摄影：陈建华 Photo by: Chen Jianhua

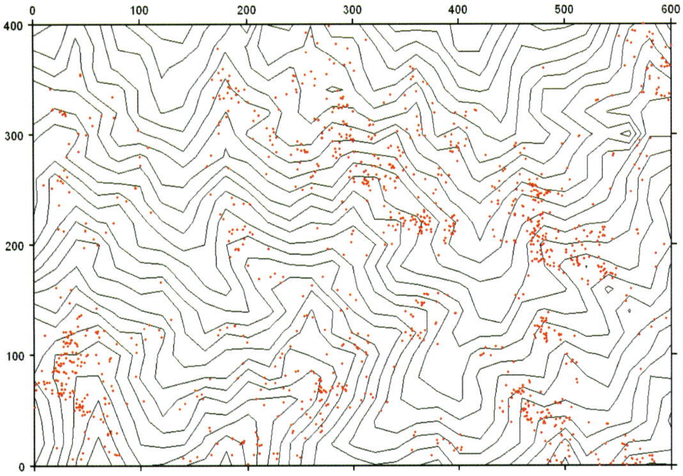

个体分布图 Distribution of Individuals

径级分布表 DBH Class

胸径区间 (Diameter class) (cm)	个体数 (No. of individuals in the plot)	比例 (Proportion)
1～2	603	0.49
2～5	333	0.27
5～10	111	0.09
10～20	111	0.09
20～30	54	0.04
30～60	19	0.02
≥60	0	0.00

9 钩栲（钩锥） gōukǎo(gōuzhuī) | Hook Oatchestnut

Castanopsis tibetana Hance
壳斗科 | Fagaceae

代码（SpCode）= CASTIB
个体数（Individual number/24ha）= 258
最大胸径（Max DBH）= 53 cm
重要值排序（Importance value rank）= 58

常绿乔木，高达30m。树皮灰褐色，粗糙。新生嫩叶暗紫褐色。成长叶革质。壳斗连刺直径6～8cm，刺长1.5～2.5cm，坚果高1.5～1.8cm，横径2～2.8cm，被毛。花期4～5月，果翌年8～10月成熟。

Evergreen tree, up to 30 m tall. Bark dark brown, rough. Leaf blade leathery, dark purple-brown when young. Cupule globose, 6-8 cm in diam, bracts spinelike, entirely covering cupule, 1.5-2.5 cm; nut 1 per cupule, 1.5-1.8 cm × 2-2.8 cm, hairy. Fl. Apr.-May, fr. Aug.-Oct. of the following year.

树皮 Bark
摄影：方腾 Photo by: Fang Teng

花序 Inflorescence
摄影：方腾 Photo by: Fang Teng

果枝 Fruiting branch
摄影：刘晓娟 Photo by: Liu Xiaojuan

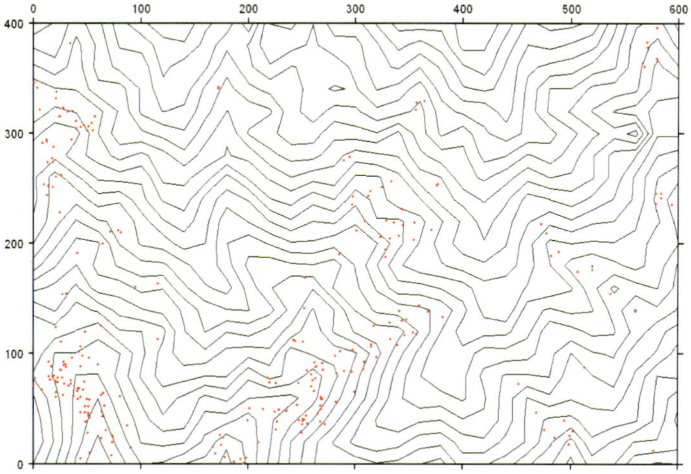

个体分布图 Distribution of Individuals

径级分布表 DBH Class

胸径区间 (Diameter class) (cm)	个体数 (No. of individuals in the plot)	比例 (Proportion)
1～2	76	0.29
2～5	75	0.29
5～10	40	0.15
10～20	39	0.15
20～30	17	0.07
30～60	12	0.05
≥60	0	0.00

10 青冈（青冈栎）

qīnggāng(qīnggānglì) | Qinggang

Cyclobalanopsis glauca (Thunb.) Oerst.
壳斗科 | Fagaceae

代码（SpCode）= CYCGLA
个体数（Individual number/24ha）= 1620
最大胸径（Max DBH）= 45 cm
重要值排序（Importance value rank）= 21

常绿乔木，高达20m。叶片革质，倒卵状椭圆形或长椭圆形，边缘中部以上有疏锯齿。壳斗碗形，包着坚果1/3～1/2。坚果椭圆形，直径约0.9～1.4cm，高1～1.6cm。花期4～5月，果期10月。

Evergreen tree, up to 20 m tall. Leaf blade leathery, obovate to oblong-elliptic, apical half of the margen remotely serrate. Cupule bowl-shaped, enclosing 1/3-1/2 of nut. Nut ellipsoid, 0.9-1.4 cm × 1-1.6 cm. Fl. Apr.-May, fr. Oct..

树皮 Bark
摄影：方腾 Photo by: Fang Teng

枝叶 Branch and leaves
摄影：张金龙 Photo by: Zhang Jinlong

果枝 Fruiting branches
摄影：方腾 Photo by: Fang Teng

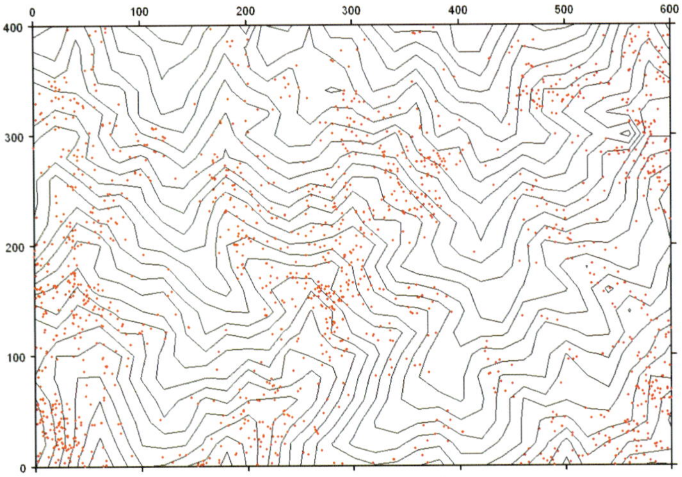

个体分布图 Distribution of Individuals

径级分布表 DBH Class

胸径区间 (Diameter class) (cm)	个体数 (No. of individuals in the plot)	比例 (Proportion)
1～2	241	0.15
2～5	328	0.20
5～10	282	0.17
10～20	592	0.37
20～30	163	0.10
30～60	14	0.01
≥60	0	0.00

11 细叶青冈（岩青冈） xìyèqīnggāng(yánqīnggāng) | Thinleaf Qinggang

Cyclobalanopsis gracilis (Rehd. et Wils.) W. C. Cheng et T. Hong
壳斗科 | Fagaceae

代码（SpCode）= CYCGRA
个体数（Individual number/24ha）= 7
最大胸径（Max DBH）= 7 cm
重要值排序（Importance value rank）= 125

常绿乔木，高达15m，树皮灰褐色。叶长卵形至卵状披针形，叶背灰绿色并有平伏单毛，上面亮绿色；叶缘1/3以上有细尖锯齿。壳斗直径1～1.3cm，高6～8mm，坚果直径约1cm，高1.5～2cm。花期3～4月，果期10～11月。

Evergreen tree, up to 15 m tall. Bark dark brownd. Leaf blade oblong-ovate to ovate-lanceolate, abaxially whitish and with prostrate simple hairs, adaxially bright green, margin apical 2/3 sharply serrulate. Cupule bowl-shaped, 6-8 mm × 1-1.3 cm. Nut ellipsoid, 1.5-2 cm × ca. 1 cm, apex hairy. Fl. Mar-Apr., fr. Oct.-Nov..

树干 Trunk
摄影：杨波 Photo by: Yang Bo

枝叶 Branch and leaves
摄影：陈彬 Photo by: Chen Bin

叶背 Leaf back
摄影：杨波 Photo by: Yang Bo

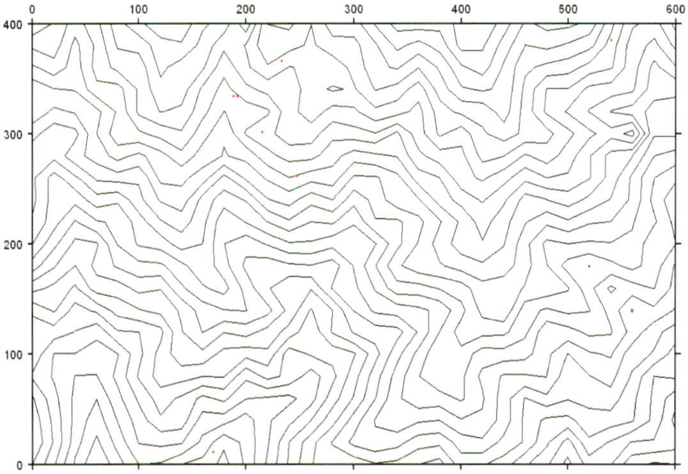

个体分布图 Distribution of Individuals

径级分布表 DBH Class

胸径区间 (Diameter class) (cm)	个体数 (No. of individuals in the plot)	比例 (Proportion)
1～2	6	0.86
2～5	0	0.00
5～10	1	0.14
10～20	0	0.00
20～30	0	0.00
30～60	0	0.00
≥60	0	0.00

12 小叶青冈（青栲） xiǎoyèqīnggāng(qīngkǎo) | Littleleaf Qinggang

Cyclobalanopsis myrsinifolia (Blume) Oerst.
壳斗科 | Fagaceae

代码（SpCode）= CYCMYR
个体数（Individual number/24ha）= 375
最大胸径（Max DBH）= 55 cm
重要值排序（Importance value rank）= 45

常绿乔木，高达20m。小枝被突起的淡褐色长圆形皮孔。叶缘中部以上有细锯齿。叶背被粉白色粉末，干后深灰色，叶缘1/2以上具锯齿。坚果卵形或椭圆形，直径1～1.5cm，高1.4～2.5cm。花期6月，果期10月。

Evergreen tree, up to 20 m tall. Branchlets lenticellate; lenticels grayish brown, oblong, convex. Leaf blade abaxially whitish farinose but dark gray when dry, adaxially green and glabrous; margin apical 1/2 serrulate. Nut ovoid to ellipsoid, 1.4-2.5 cm × 1-1.5 cm. Fl. Jun., fr. Oct..

树干 Trunk
摄影：方腾 Photo by: Fang Teng

枝叶 Branch and leaves
摄影：方腾 Photo by: Fang Teng

叶背 Leaf back
摄影：方腾 Photo by: Fang Teng

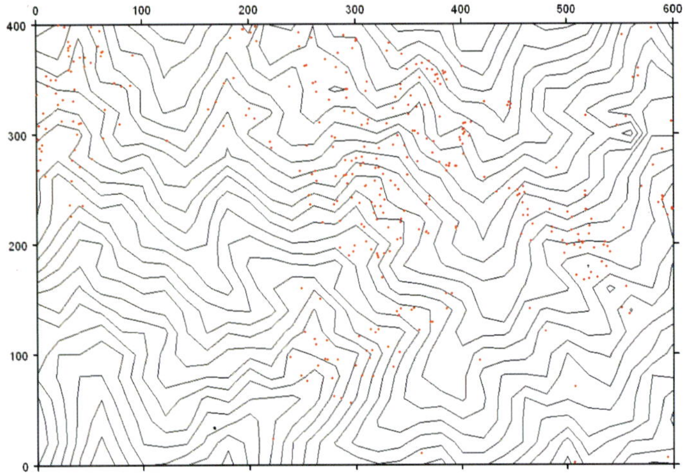

个体分布图 Distribution of Individuals

径级分布表 DBH Class

胸径区间 (Diameter class) (cm)	个体数 (No. of individuals in the plot)	比例 (Proportion)
1～2	92	0.25
2～5	39	0.10
5～10	19	0.05
10～20	64	0.17
20～30	80	0.21
30～60	81	0.22
≥60	0	0.00

13 柯（石栎、石栗） kē(shílì、shílì) | Tanoak

Lithocarpus glaber (Thunb.) Nakai
壳斗科 | Fagaceae

代码（SpCode）= LITGLA
个体数（Individual number/24ha）= 1313
最大胸径（Max DBH）= 38.6 cm
重要值排序（Importance value rank）= 20

常绿乔木，高达15m。一年生枝密被灰黄色短绒毛。叶革质或厚纸质，上部有2～4个浅裂齿或全缘。叶背有较厚的蜡鳞层，指甲刮过会有明显白痕。花期7～11月，果翌年同期成熟。

Evergreen tree, up to 15 m tall; branchlets and rachis of inflorescences densely tawny tomentose. Leaf blade leathery to thickly papery, abaxially with dense scalelike glands, whitish mark conspicuous if scratched by nail; margin entire or with 2-4 teeth on apical part. Fl. Jul.-Nov., fr. Jul.-Nov. of the following year.

树干 Trunk
摄影：方腾 Photo by: Fang Teng

叶 Leaf
摄影：张金龙 Photo by: Zhang Jinlong

叶背 Leaf back
摄影：张金龙 Photo by: Zhang Jinlong

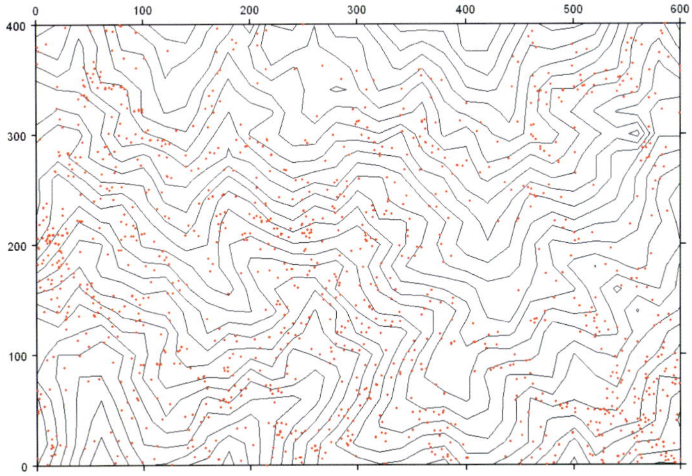

个体分布图 Distribution of Individuals

径级分布表 DBH Class

胸径区间 (Diameter class) (cm)	个体数 (No. of individuals in the plot)	比例 (Proportion)
1～2	343	0.26
2～5	339	0.26
5～10	222	0.17
10～20	302	0.23
20～30	98	0.07
30～60	9	0.01
≥60	0	0.00

14 枹栎（短柄枹） bāolì(duǎnbǐngbāo) | Serrate Oak

Quercus serrata Murray

壳斗科 | Fagaceae

代码（SpCode）= QUESER

个体数（Individual number/24ha）= 3508

最大胸径（Max DBH）= 42 cm

重要值排序（Importance value rank）= 25

落叶乔木，高达25m。树皮灰褐色，深纵裂。幼枝被柔毛，不久即脱落。叶片薄革质，倒卵形或倒卵状椭圆形，叶缘有腺状锯齿。花期3～4月，果期9～10月。

Deciduous tree, up to 25 m tall. Bark gray-brown, deeply longitudinal crack. Leaves subsessile to petiolate; petiole to 3 cm, glabrous or glabrescent; leaf blade thinly leathery, narrowly elliptic-ovate, ovate-lanceolate, or obovate, margin glandular serrate. Fl. Mar.-Apr., fr. Sep.-Oct..

雄花序 Male inflorescence

摄影：陈彬 Photo by: Chen Bin

幼苗 Seedling

摄影：金冬梅 Photo by: Jin Dongmei

果枝 Fruiting branch

摄影：张金龙 Photo by: Zhang Jinlong

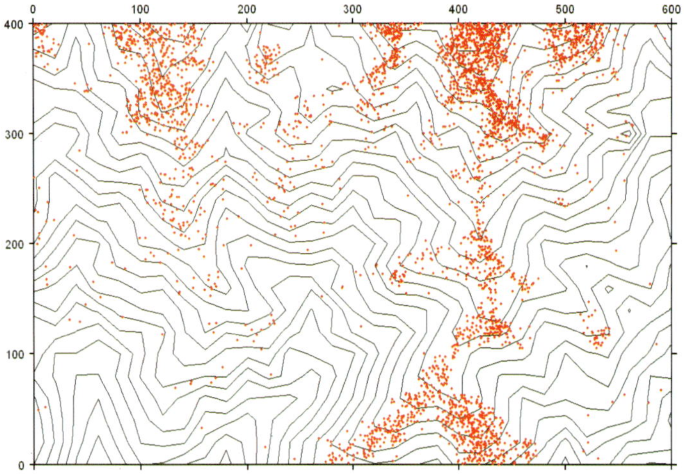

个体分布图 Distribution of Individuals

径级分布表 DBH Class

胸径区间 (Diameter class) (cm)	个体数 (No. of individuals in the plot)	比例 (Proportion)
1～2	76	0.02
2～5	529	0.15
5～10	1204	0.34
10～20	1500	0.43
20～30	192	0.05
30～60	7	0.00
≥60	0	0.00

15 乌冈栎 wūgānglì | Wugang Oak

Quercus phillyreoides A. Gray
壳斗科 | Fagaceae

代码（SpCode）= QUEPHI
个体数（Individual number/24ha）= 10
最大胸径（Max DBH）= 23.5 cm
重要值排序（Importance value rank）= 121

常绿灌木或小乔木。小枝纤细，灰褐色。叶片革质，较小，倒卵形或窄椭圆形，长2～8cm，宽1.5～3cm，叶缘有细密锯齿。壳斗杯形，包被坚果1/3～1/2。坚果椭圆形，直径0.5～1 cm，高1～2 cm。花期3～4月，果期9～10月。

Evergreen shrub or small tree, up to 10 m tall. Branchlets slender, grayish brown; leaf blade small, obovate, narrowly elliptic, or narrowly ovate, 2-8 cm × 1.5-3 cm, margin densely with small teeth. Cupule cupular, enclosing 1/3-1/2 of nut; nut ellipsoid, 0.5-1 cm × 1-2 cm. Fl. Mar.-Apr., fr. Sep.-Oct..

树干 Trunk
摄影：方腾 Photo by: Fang Teng

枝叶 Branch and leaves
摄影：张金龙 Photo by: Zhang Jinlong

坚果与总苞 Nut and involucre
摄影：陈国科 Photo by: Chen Guoke

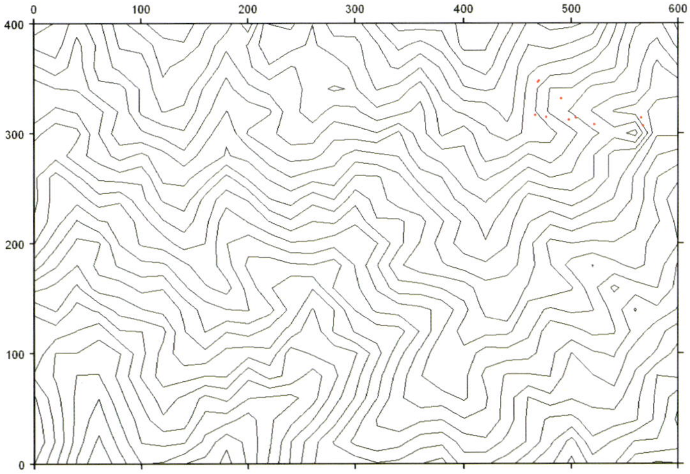

个体分布图 Distribution of Individuals

径级分布表 DBH Class

胸径区间 (Diameter class) (cm)	个体数 (No. of individuals in the plot)	比例 (Proportion)
1～2	0	0.00
2～5	1	0.10
5～10	4	0.40
10～20	4	0.40
20～30	1	0.10
30～60	0	0.00
≥60	0	0.00

16 紫弹树（黄果朴） zǐdànshù(huángguǒpǔ) | Biond Nettletree

Celtis biondii Pamp.
榆科 | Ulmaceae

代码（SpCode）= CELBIO
个体数（Individual number/24ha）= 24
最大胸径（Max DBH）= 41.8 cm
重要值排序（Importance value rank）= 95

落叶小乔木至乔木，高达18m。树皮暗灰色。叶宽卵形、卵形至卵状椭圆形，在中部以上疏具浅齿。每叶腋有果1～3枚，核果黄色至橘红色。花期4～5月，果期9～10月。

Deciduous tree, up to 18 m tall. Bark dark gray. Leaf blade broadly ovate, ovate, or ovate-elliptic, margin shallowly serrate on apical half. Infructescences 1-3 per leaf axil. Drupe yellow to reddish orange. Fl. Apr.-May, fr. Sep.-Oct..

幼苗 Seedling
摄影：金冬梅 Photo by: Jin Dongmei

枝叶 Branch and leaves
摄影：陈建华 Photo by: Chen Jianhua

核果 Drupe
摄影：方腾 Photo by: Fang Teng

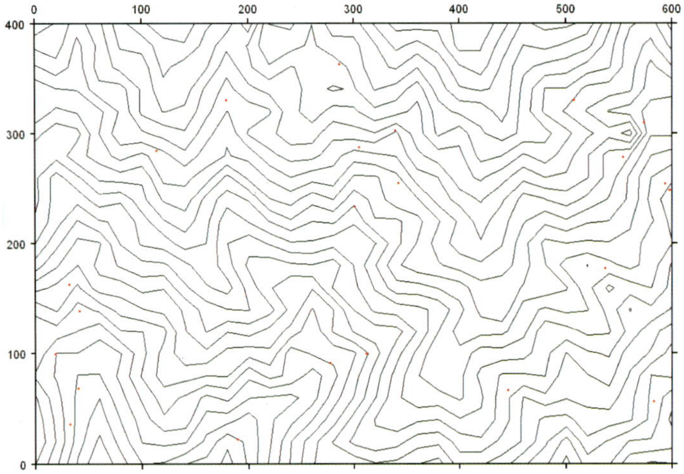

个体分布图 Distribution of Individuals

径级分布表 DBH Class

胸径区间 (Diameter class) (cm)	个体数 (No. of individuals in the plot)	比例 (Proportion)
1～2	8	0.33
2～5	5	0.21
5～10	4	0.17
10～20	4	0.17
20～30	2	0.08
30～60	1	0.04
≥60	0	0.00

17 构棘（葨芝） gòují(wēizhī) | Vietnam Cudrania

Maclura cochinchinensis (Lour.) Corner
桑科 | Moraceae

代码（SpCode）= MACCOC
个体数（Individual number/24ha）= 7
最大胸径（Max DBH）= 3.6 cm
重要值排序（Importance value rank）= 126

直立或攀缘状常绿灌木。枝无毛，具粗壮直立或略弯的腋生刺。叶纸质到革质，椭圆状披针形或长圆形，全缘。聚合果肉质，直径2～5cm，成熟时橙红色。花期4～5月，果期6～7月。

Evergreen shrub, erect or scandent. Branches glabrous, spines stout, straight or little curved. Leaf blade elliptic-lanceolate to oblong, papery to leathery. Fruiting syncarp fleshy, reddish orange when mature, 2-5 cm in diam. Fl. Apr.-May, fr. Jun.-Jul..

幼苗 Seedling
摄影：金冬梅 Photo by: Jin Dongmei

枝叶 Branch and leaves
摄影：陈建华 Photo by: Chen Jianhua

聚花果 Multiple fruit
摄影：陈建华 Photo by: Chen Jianhua

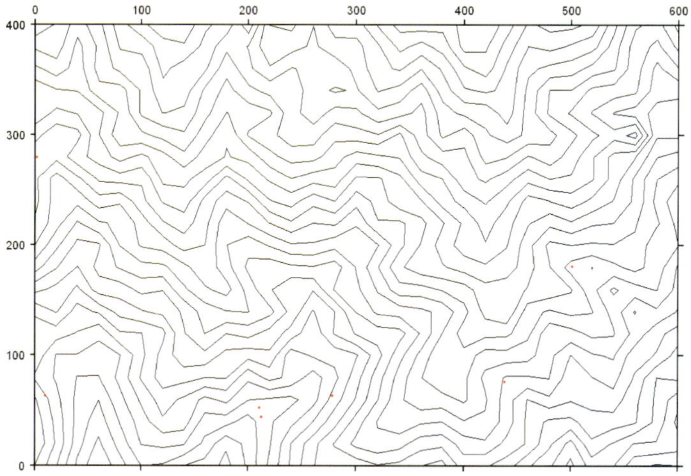

个体分布图 Distribution of Individuals

径级分布表 DBH Class

胸径区间 (Diameter class) (cm)	个体数 (No. of individuals in the plot)	比例 (Proportion)
1～2	5	0.71
2～5	2	0.29
5～10	0	0.00
10～20	0	0.00
20～30	0	0.00
30～60	0	0.00
≥60	0	0.00

18 天仙果

tiānxiānguǒ | Fairy Fig

Ficus erecta Thunb.

桑科 | Moraceae

代码（SpCode）= FICERE

个体数（Individual number/24ha）= 7

最大胸径（Max DBH）= 4.6 cm

重要值排序（Importance value rank）= 118

落叶或半常绿小乔木或灌木，高2～7m；树皮灰褐色，小枝光滑或密生硬毛。叶厚纸质。托叶三角状披针形，膜质，早落。榕果单生叶腋，梨形或球形，直径1.2～2cm。花果期5～6月。

Shrub or small tree, 2-7 m tall, deciduous or semideciduous. Bark grayish brown. Branchlets glabrous or densely brown tomentose. Stipules caducous, reddish brown, broadly ovate or triangular-lanceolate; leaf blade thickly papery. Figs axillary on normal leafy shoots, solitary, globose to pear-shaped, 1.2-2 cm in diam. Fl. and fr. May-Jun..

榕果 Multiple fruit

摄影：陈国科 Photo by: Chen Guoke

幼苗 Seedling

摄影：金冬梅 Photo by: Jin Dongmei

枝叶 Branch and leaves

摄影：方腾 Photo by: Fang Teng

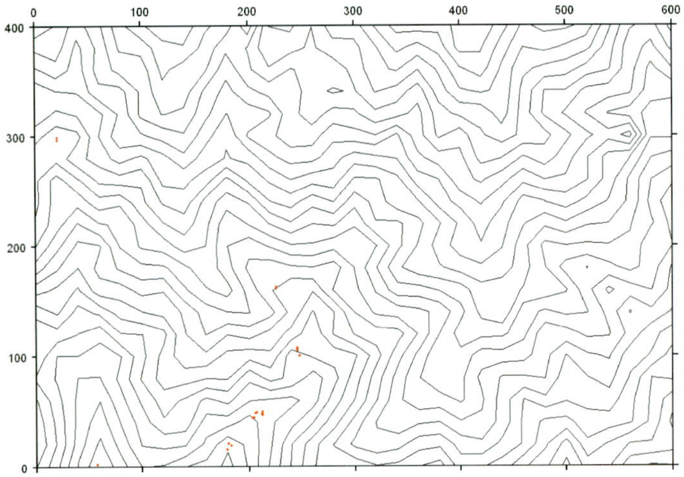

个体分布图 Distribution of Individuals

径级分布表 DBH Class

胸径区间 (Diameter class) (cm)	个体数 (No. of individuals in the plot)	比例 (Proportion)
1～2	15	0.63
2～5	9	0.38
5～10	0	0.00
10～20	0	0.00
20～30	0	0.00
30～60	0	0.00
≥60	0	0.00

19 阔叶十大功劳 kuòyèshídàgōngláo | Broadleaf Mahonia

Mahonia bealei (Fort.) Carr.
小檗科 | Berberidaceae

代码（SpCode）= MAHBEA
个体数（Individual number/24ha）= 3
最大胸径（Max DBH）= 1.1 cm
重要值排序（Importance value rank）= 139

常绿灌木，高1～2m。茎单生，无侧枝。小叶厚革质。浆果深蓝色，被白粉，短圆形，多汁。花期9月到翌年1月，果期翌年3～5月。

Evergreen shrub, 1-2 m tall. Stems monomorphic, without short axillary shoots. Leaflet blades thick and rigid. Berries dark blue, glaucous, oblong-ovoid, juicy. Fl. Sep. to Jan. of the the following year, fr. Mar.-May of the second year.

叶 Leaf
摄影：陈彬 Photo by: Chen Bin

花序 Inflorescence
摄影：杨波 Photo by: Yang Bo

果序 Infructescence
摄影：陈彬 Photo by: Chen Bin

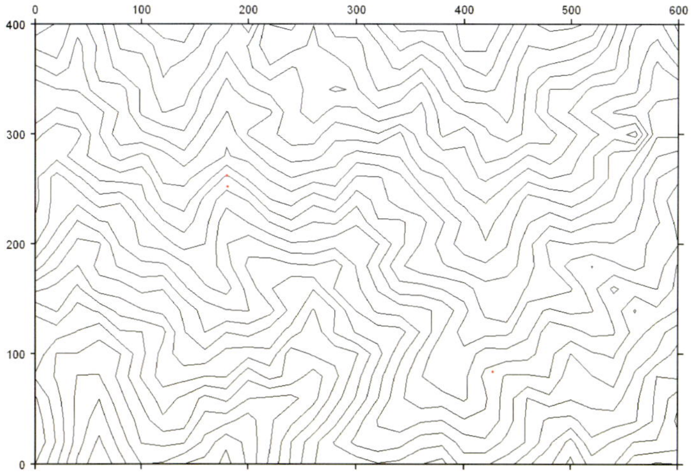

个体分布图 Distribution of Individuals

径级分布表 DBH Class

胸径区间 (Diameter class) (cm)	个体数 (No. of individuals in the plot)	比例 (Proportion)
1～2	3	1.00
2～5	0	0.00
5～10	0	0.00
10～20	0	0.00
20～30	0	0.00
30～60	0	0.00
≥60	0	0.00

20 黄山木兰

huángshānmùlán | Huangshan Magnolia

Magnolia cylindrica Wils.

木兰科 | Magnoliaceae

代码（SpCode）= MAGCYL

个体数（Individual number/24ha）= 16

最大胸径（Max DBH）= 26 cm

重要值排序（Importance value rank）= 108

落叶乔木，高达10m，树皮灰白色，平滑。嫩枝、叶柄、叶背被淡黄色平伏毛。老枝紫褐色，皮揉碎有辛辣香气。花先叶开放。花期2～3月，果期8～9月。

Deciduous tree, up to 10 m tall. Bark grayish white, smooth. Young twigs, petioles, and leaf blade abaxial surfaces with pale yellow appressed trichomes. Old twigs purplish brown, bark with pungent odor when crushed. Flowers appearing before leaves. Fl. Feb.-Mar., fr. Aug.-Sep..

花 Flower

摄影：杨波 Photo by: Yang Bo

花芽 Flower-bud

摄影：杨波 Photo by: Yang Bo

果枝 Fruiting branch

摄影：陈彬 Photo by: Chen Bin

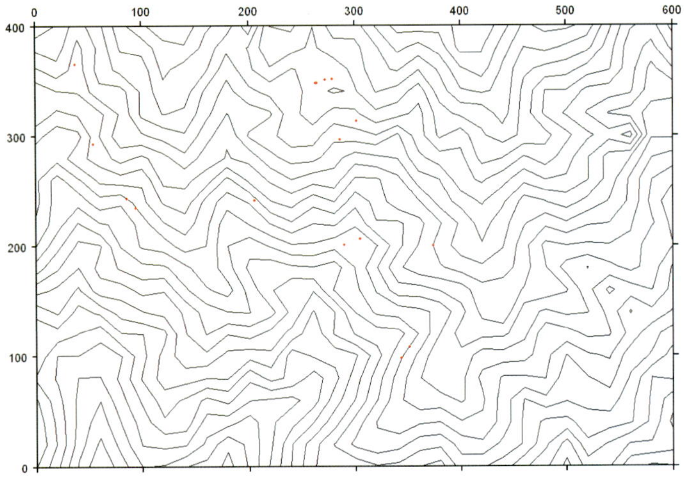

个体分布图 Distribution of Individuals

径级分布表 DBH Class

胸径区间 (Diameter class) (cm)	个体数 (No. of individuals in the plot)	比例 (Proportion)
1～2	0	0.00
2～5	3	0.19
5～10	2	0.13
10～20	6	0.38
20～30	5	0.31
30～60	0	0.00
≥60	0	0.00

21 木莲（乳源木莲） mùlián(rǔyuánmùlián) | Woodlotus

Manglietia fordiana Oliv.
木兰科 | Magnoliaceae

代码（SpCode）= MANFOR
个体数（Individual number/24ha）= 16
最大胸径（Max DBH）= 43 cm
重要值排序（Importance value rank）= 102

常绿乔木，高达20m。树皮灰褐色。嫩枝及营养芽有红褐色柔毛，后脱落无毛。叶片革质或薄革质。单花顶生，花被9，3轮，白色。花期4～5月，果期9～10月。

Evergreen tree, up to 20 m tall. Bark pale grayish brown. Young twigs and vegetative buds reddish brown pubescent, later glabrescent. Leaf blade leathery or thinly leathery. Flower solitary on twig apex. Tepals 9, 3 per whorl, white. Fl. Apr.-May, fr. Sep.-Oct..

树干 Trunk
摄影：张金龙 Photo by: Zhang Jinlong

花 Flower
摄影：金冬梅 Photo by: Jin Dongmei

蓇葖果 Follicle
摄影：张金龙 Photo by: Zhang Jinlong

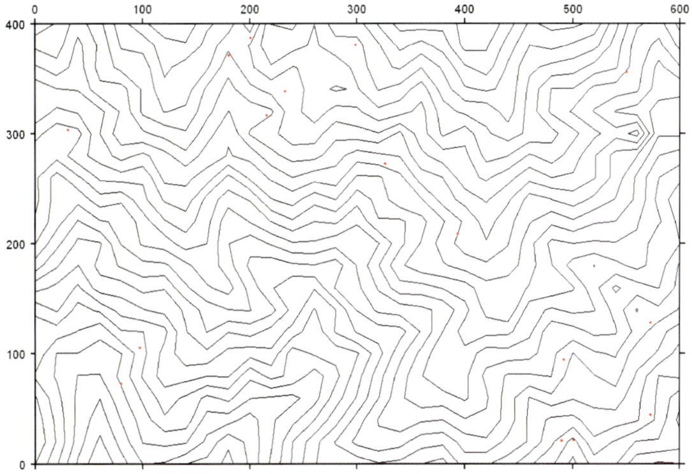

个体分布图 Distribution of Individuals

径级分布表 DBH Class

胸径区间 (Diameter class) (cm)	个体数 (No. of individuals in the plot)	比例 (Proportion)
1～2	2	0.13
2～5	0	0.00
5～10	0	0.00
10～20	2	0.13
20～30	10	0.63
30～60	2	0.13
≥60	0	0.00

22 野含笑

yěhánxiào | Wild Michelia

Michelia skinneriana Dunn
木兰科 | Magnoliaceae

代码（SpCode）= MICSKI
个体数（Individual number/24ha）= 243
最大胸径（Max DBH）= 24.8 cm
重要值排序（Importance value rank）= 70

常绿乔木，高可达15m，树皮灰白色，平滑。芽、嫩枝、叶柄、叶背中脉及花梗均密被褐色长柔毛。叶革质。花梗纤细，花被6片，淡黄色。花期5～6月，果期8～9月。

Evergreen tree, up to 15 m tall. Bark grayish white, smooth. Young twigs, buds, petioles, leaf blade midrib abaxially, and peduncles densely brown villous. Leaf blade leathery. Peduncle slender. Tepals 6, pale yellow. Fl. May-Jun., fr. Aug.-Sep..

花枝 Flowering branch
摄影：方腾 Photo by: Fang Teng

叶 Leaf
摄影：张金龙 Photo by: Zhang Jinlong

蓇葖果 Follicle
摄影：张金龙 Photo by: Zhang Jinlong

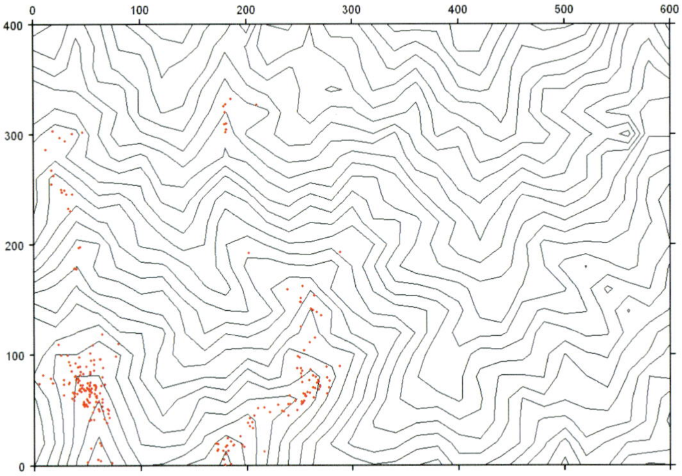

个体分布图 Distribution of Individuals

径级分布表 DBH Class

胸径区间 (Diameter class) (cm)	个体数 (No. of individuals in the plot)	比例 (Proportion)
1～2	83	0.34
2～5	96	0.40
5～10	43	0.18
10～20	20	0.08
20～30	1	0.00
30～60	0	0.00
≥60	0	0.00

23 红毒茴（披针叶茴香、莽草） hóngdúhuí(pīzhēnyèhuíxiāng、mǎngcǎo) | Poisonous Eightangle

Illicium lanceolatum A. C. Smith
木兰科 | Magnoliaceae

代码（SpCode）= ILLLAN
个体数（Individual number/24ha）= 51
最大胸径（Max DBH）= 14 cm
重要值排序（Importance value rank）= 117

常绿灌木或小乔木，高 3 ～ 10m。树皮浅灰色至灰褐色。叶革质，集生枝顶。花被片 10 ～ 15，红色，肉质。果实具有 9 ～ 14 个蓇葖，果形似八角，有毒，不可食用。花期 4 ～ 6 月，果期 9 ～ 10 月。

Evergreen shrub or small tree, up to 10 m tall. Bark grayish white to greyish brown. Leaves leathery, in clusters at distal nodes. Tepals 10-15, red to dark red, fleshy. Fruit with 9-14 follicles, likes star anise but poisonous, unedible. Fl. Apr.-Jun., fr. Sep.-Oct..

蓇葖果 Follicle
摄影：方腾 Photo by: Fang Teng

幼苗 Seedling
摄影：金冬梅 Photo by: Jin Dongmei

花枝 Flowering branch
摄影：金冬梅 Photo by: Jin Dongmei

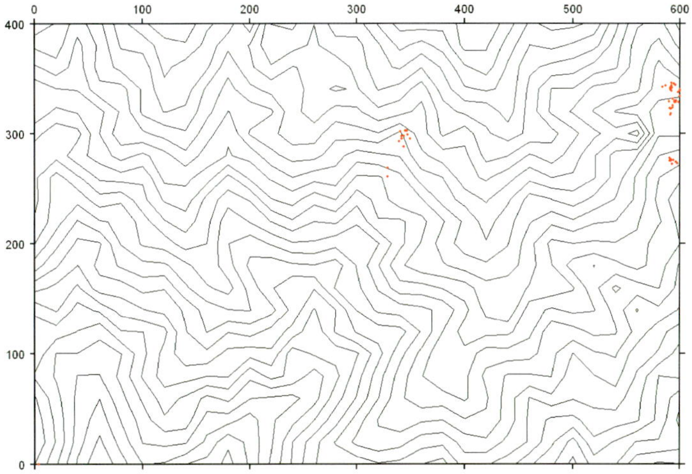

个体分布图 Distribution of Individuals

径级分布表 DBH Class

胸径区间 (Diameter class) (cm)	个体数 (No. of individuals in the plot)	比例 (Proportion)
1～2	15	0.29
2～5	28	0.54
5～10	6	0.12
10～20	3	0.06
20～30	0	0.00
30～60	0	0.00
≥60	0	0.00

24 柳叶蜡梅 liǔyèlàméi | Willowleaf Wintersweet

Chimonanthus salicifolius S. Y. Hu
蜡梅科 | Calycanthaceae

代码（SpCode）= CHISAL
个体数（Individual number/24ha）= 7835
最大胸径（Max DBH）= 19.5 cm
重要值排序（Importance value rank）= 6

半常绿灌木。小枝嫩时有四棱，老后变圆筒形，多毛。叶近革质，上表面粗糙，背面被不明显的短柔毛。花期8～10月，果期翌年5月。

Semi-evergreen shrub. Branchlets quadrangular when young but becoming subterete, hirsute. Leaf blade subleathery, both surfaces scabrous, abaxially hispidulous on veins and margin, adaxially light grayish green and with inconspicuous pubescence. Fl. Aug.-Oct., fr. May of the following year.

幼苗 Seedling
摄影：金冬梅 Photo by: Jin Dongmei

枝叶 Branch and leaves
摄影：陈建华 Photo by: Chen Jianhua

花枝 Flowering branch
摄影：张金龙 Photo by: Zhang Jinlong

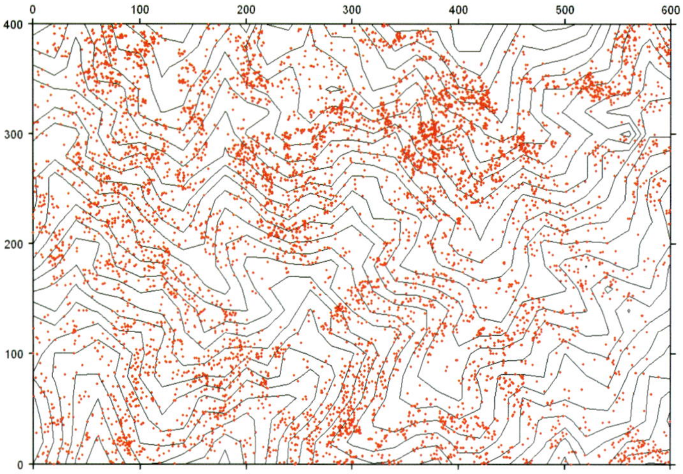

个体分布图 Distribution of Individuals

径级分布表 DBH Class

胸径区间 (Diameter class) (cm)	个体数 (No. of individuals in the plot)	比例 (Proportion)
1～2	5802	0.74
2～5	2026	0.26
5～10	5	0.00
10～20	2	0.00
20～30	0	0.00
30～60	0	0.00
≥60	0	0.00

25 天竺桂（浙江樟） tiānzhúguì(zhèjiāngzhāng) | Zhejiang Cinnamon

Cinnamomum japonicum Siebold
樟科 | Lauraceae

代码（SpCode）= CINJAP
个体数（Individual number/24ha）= 319
最大胸径（Max DBH）= 26.5 cm
重要值排序（Importance value rank）= 54

常绿乔木，高达15m。小枝绿色，光滑，有香气。叶近对生，或者在小枝上部互生，革质，两面光滑，离基三出脉。花期4～5月，果期10～11月。

Evergreen tree, up to 15 m tall. Branchlets green, very glabrous, scented. Leaves subopposite or those on upper part of branchlet alternate; leaf blade leathery, glabrous on both surfaces, triplinerved. Fl. Apr.-May, fr. Oct.-Nov..

树皮 Bark
摄影：方腾 Photo by: Fang Teng

枝叶 Branch and leaves
摄影：张金龙 Photo by: Zhang Jinlong

核果 Drupe
摄影：陈彬 Photo by: Chen Bin

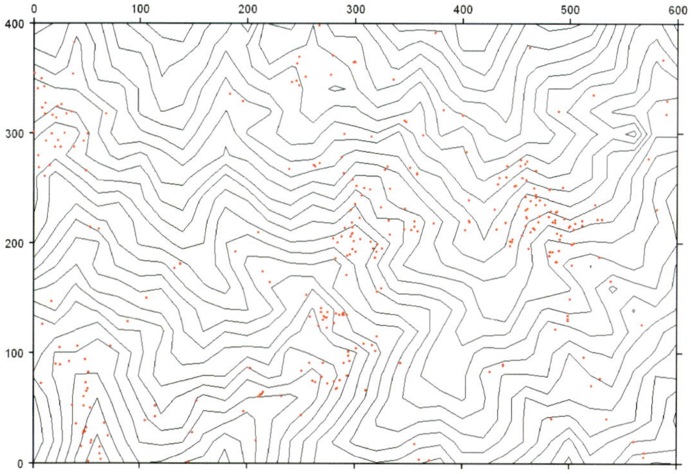

个体分布图 Distribution of Individuals

径级分布表 DBH Class

胸径区间 (Diameter class) (cm)	个体数 (No. of individuals in the plot)	比例 (Proportion)
1～2	159	0.50
2～5	118	0.37
5～10	36	0.11
10～20	4	0.01
20～30	2	0.01
30～60	0	0.00
≥60	0	0.00

26 香桂（细叶香桂） xiāngguì(xìyèxiāngguì) | Fragrant Cinnamon

Cinnamomum subavenium Miq.
樟科 | Lauraceae

代码（SpCode）= CINSUB
个体数（Individual number/24ha）= 1958
最大胸径（Max DBH）= 40.7 cm
重要值排序（Importance value rank）= 18

常绿乔木，高达20m。树皮黄色，光滑。小枝纤细，密被黄色平伏绢状短柔毛。叶革质，嫩时有贴伏的黄色绢毛，三出脉或近离基三出脉。花期6～7月，果期8～10月。

Evergreen tree, up to 20 m tall. Bark gray, smooth. Branchlets slender, densely covered with yellow appressed sericeous-pubescent hairs. Leaf blade leathery, densely appressed yellow sericeous hair when young but glabrate to glabrous when mature, trinerved or triplinerved. Fl. Jun.-Jul., fr. Aug.-Oct..

幼苗 Seedling
摄影：金冬梅 Photo by: Jin Dongmei

叶 Leaf
摄影：张金龙 Photo by: Zhang Jinlong

叶背 Leaf back
摄影：陈国科 Photo by: Chen Guoke

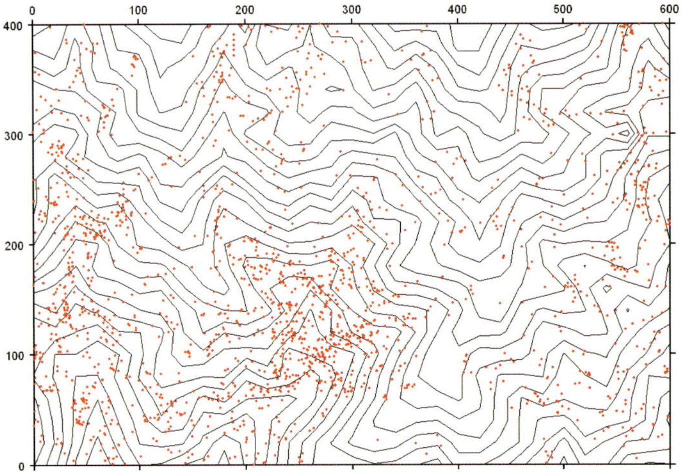

个体分布图 Distribution of Individuals

径级分布表 DBH Class

胸径区间 (Diameter class) (cm)	个体数 (No. of individuals in the plot)	比例 (Proportion)
1～2	887	0.45
2～5	823	0.42
5～10	224	0.11
10～20	20	0.01
20～30	2	0.00
30～60	2	0.00
≥60	0	0.00

27 乌药 wūyào | Combined Spicebush

Lindera aggregata (Sims) Kosterm.
樟科 | Lauraceae

代码（SpCode）= LINAGG
个体数（Individual number/24ha）= 13
最大胸径（Max DBH）= 2 cm
重要值排序（Importance value rank）= 110

常绿灌木或小乔木，高达5m。树皮灰褐色，根有纺锤状或结节状膨大。叶互生，革质，三出脉。花期3～4月，果期5～11月。

Evergreen shrub or small tree, up to 5 m tall. Bark greyish brown; roots fusiform-inflated or nodose-inflated. Leaves alternate; leaf blade leathery or subleathery, trinerved. Fl. Mar.-Apr., fr. May-Nov..

花枝 Flowering branch
摄影：陈彬 Photo by: Chen Bin

叶背 Leaf back
摄影：陈国科 Photo by: Chen Guoke

果枝 Fruiting branch
摄影：方腾 Photo by: Fang Teng

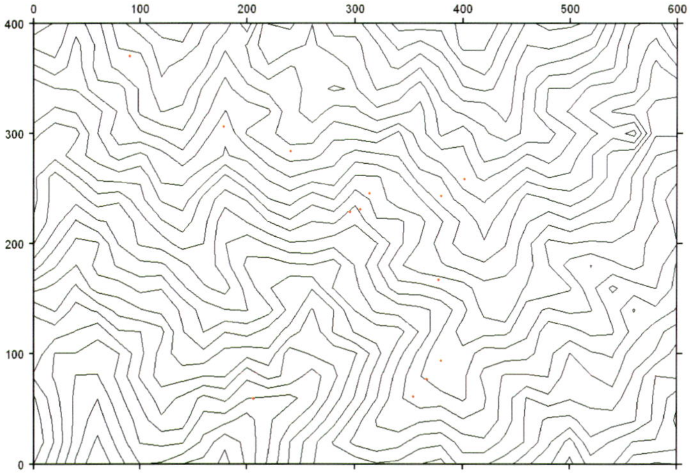

个体分布图 Distribution of Individuals

径级分布表 DBH Class

胸径区间 (Diameter class) (cm)	个体数 (No. of individuals in the plot)	比例 (Proportion)
1～2	13	1.00
2～5	0	0.00
5～10	0	0.00
10～20	0	0.00
20～30	0	0.00
30～60	0	0.00
≥60	0	0.00

28 山胡椒 shānhújiāo | Greyblue Spicebush

Lindera glauca (Sieb. et Zucc.) Blume
樟科 | Lauraceae

代码（SpCode）= LINGLA
个体数（Individual number/24ha）= 34
最大胸径（Max DBH）= 6.2 cm
重要值排序（Importance value rank）= 90

落叶灌木或小乔木，高达8m。树皮平滑，灰色或灰白色。叶互生，叶片纸质，网状脉，侧脉4～6对。叶冬天干枯而不凋落。聚伞花序腋生。核果，熟时黑褐色。花期3～4月，果期7～8月。

Deciduous shrub or small tree, up to 8 m tall. Bark smooth, gray or gray-white. Leaves alternate; leaf blade papery, pinninerved, lateral veins 4-6 pairs. Leaves die but not drop off in winter. Umbels axillary. Fruit dark brown when matured. Fl. Mar.-Apr., fr. Jul.-Aug..

果枝 Fruiting branch
摄影：方腾 Photo by: Fang Teng

叶背 Leaf back
摄影：张金龙 Photo by: Zhang Jinlong

叶冬态 Leaves in winter
摄影：方腾 Photo by: Fang Teng

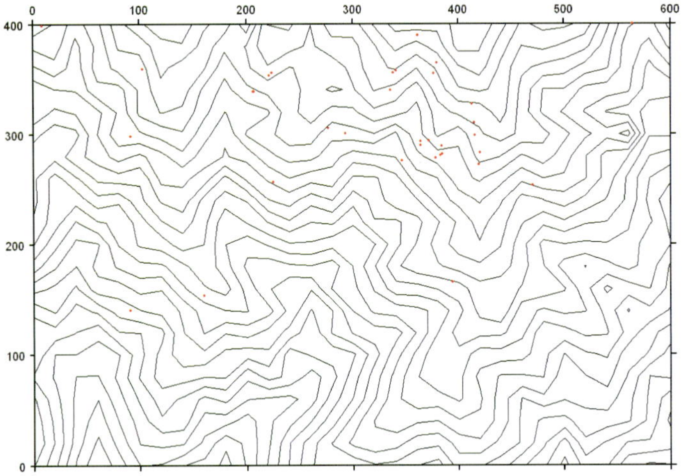

个体分布图 Distribution of Individuals

径级分布表 DBH Class

胸径区间 (Diameter class) (cm)	个体数 (No. of individuals in the plot)	比例 (Proportion)
1～2	8	0.24
2～5	24	0.71
5～10	2	0.06
10～20	0	0.00
20～30	0	0.00
30～60	0	0.00
≥60	0	0.00

29 山橿　shānjiāng | Mountain Spicebush

Lindera reflexa Hemsl.
樟科 | Lauraceae

代码（SpCode）= LINREF
个体数（Individual number/24ha）= 53
最大胸径（Max DBH）= 5 cm
重要值排序（Importance value rank）= 82

落叶灌木或小乔木。树皮棕褐色，有纵裂及斑点。幼枝黄绿色，光滑，无皮孔。叶纸质，羽状脉，侧脉6～10对。核果球形，熟时红色。花期4月，果期8月。

Deciduous shrub or small tree. Bark brown, with longitudinal splits and pits. Young branchlets yellow-green, smooth, non-lenticellate. Leaf blade papery, pinninerved, lateral veins 6-10 pairs. Fruits globose, red at maturity. Fl. Apr., fr. Aug..

核果　Drupe
摄影：方腾　Photo by: Fang Teng

幼苗　Seedling
摄影：金冬梅　Photo by: Jin Dongmei

花序　Inflorescence
摄影：方腾　Photo by: Fang Teng

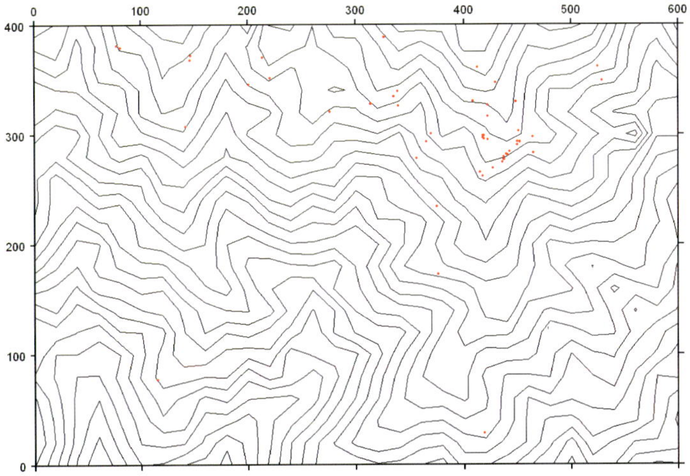

个体分布图 Distribution of Individuals

径级分布表 DBH Class

胸径区间 (Diameter class) (cm)	个体数 (No. of individuals in the plot)	比例 (Proportion)
1～2	52	0.98
2～5	1	0.02
5～10	0	0.00
10～20	0	0.00
20～30	0	0.00
30～60	0	0.00
≥60	0	0.00

30 豹皮樟 bàopízhāng | Leopard Skin Litsea

Litsea coreana Lévl. var. *sinensis* (C. K. Allen) Y. C. Yang et P. H. Huang
樟科 | Lauraceae

代码（SpCode）= LITCOR
个体数（Individual number/24ha）= 585
最大胸径（Max DBH）= 19.6 cm
重要值排序（Importance value rank）= 32

常绿小乔木。树皮呈小鳞片状剥落，剥落后呈豹皮状斑痕。叶革质，侧脉7～10条。核果近球形，果熟时由红色变黑色。花期8～9月，果期翌年夏季。

Evergreen tree. Bark gray, scaly peeled off, with collated spots exposed. Leaves alternate; leaf blade leathery, lateral veins 7-10 pairs. Fruit subglobose, from red to black at maturity. Fl. Aug.-Sep., fr. Summer of next year.

树皮 Bark
摄影：张金龙 Photo by: Zhang Jinlong

枝叶 Branch and leaves
摄影：张金龙 Photo by: Zhang Jinlong

核果 Drupe
摄影：方腾 Photo by: Fang Teng

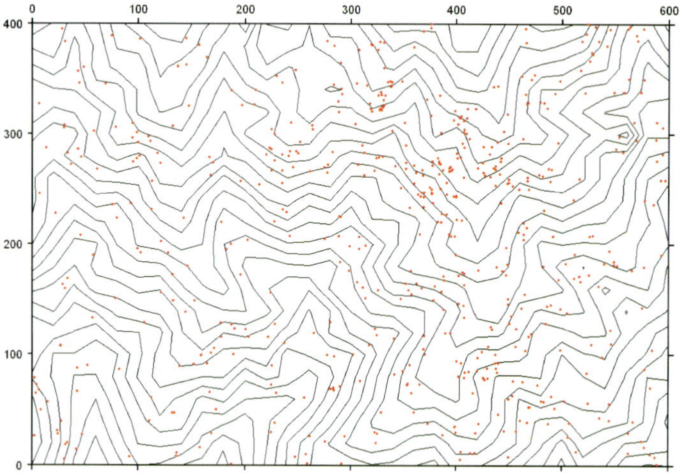

个体分布图 Distribution of Individuals

径级分布表 DBH Class

胸径区间 (Diameter class) (cm)	个体数 (No. of individuals in the plot)	比例 (Proportion)
1～2	204	0.35
2～5	267	0.46
5～10	93	0.16
10～20	21	0.04
20～30	0	0.00
30～60	0	0.00
⩾60	0	0.00

31 山鸡椒（山苍子） shānjījiāo(shāncāngzǐ) | Mountain Spicy Tree

Litsea cubeba Pers.
樟科 | Lauraceae

代码（SpCode）= LITCUB
个体数（Individual number/24ha）= 17
最大胸径（Max DBH）= 10 cm
重要值排序（Importance value rank）= 105

落叶灌木或小乔木，高达8～10m，味道芳香。小枝细长，绿色，无毛。叶互生，光滑，侧脉6～16对。核果近球形，熟时黑色。花期2～3月，果期7～8月。

Deciduous shrub or small tree, 8-10 m tall, aromatic. Branchlets slender, glabrous. Leaves alternate, glabrous, lateral veins 6-16 pairs. Fruit subglobose, black at maturity; fruiting pedicel 2-4 mm. Fl. Feb.-Mar., fr. Jul.-Aug..

果枝 Fruiting branch
摄影：方腾 Photo by: Fang Teng

幼苗 Seedling
摄影：金冬梅 Photo by: Jin Dongmei

花序 Inflorescence
摄影：方腾 Photo by: Fang Teng

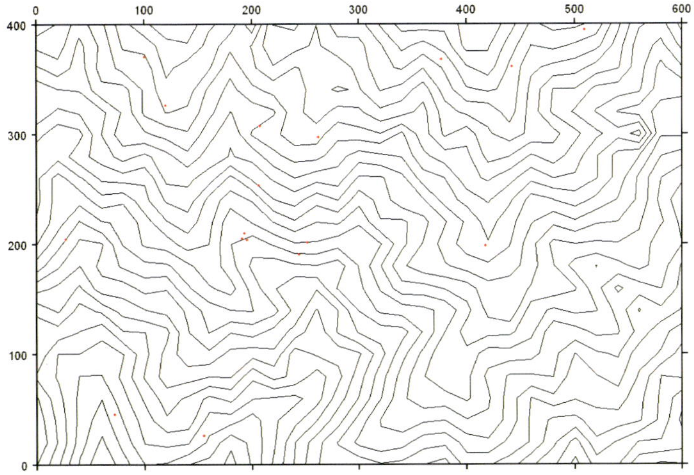

个体分布图 Distribution of Individuals

径级分布表 DBH Class

胸径区间 (Diameter class) (cm)	个体数 (No. of individuals in the plot)	比例 (Proportion)
1～2	6	0.35
2～5	5	0.29
5～10	6	0.35
10～20	0	0.00
20～30	0	0.00
30～60	0	0.00
≥60	0	0.00

32 黄丹木姜子

huángdānmùjiāngzǐ | Elongate Litsea

Litsea elongata (Wall. ex Ness) Benth. et Hook. f.

樟科 | Lauraceae

代码（SpCode）= LITELO

个体数（Individual number/24ha）= 44

最大胸径（Max DBH）= 18.3 cm

重要值排序（Importance value rank）= 94

常绿小乔木或中乔木，高达12m。树皮灰黄色或褐色，小枝密被褐色绒毛。叶互生或近轮生，密被褐色毛。果托杯状，核果长圆形，成熟时黑紫色。花期5～11月，果期翌年2～6月。

Evergreen small or medium-sized tree, up to 12 m tall. Bark greyish yellow or brown. Branchlets densely brown tomentose. Leaves alternate or subverticillate; densely brown tomentose. Fruit oblong, seated on cup-shaped perianth tube, black-purple at maturity. Fl. May-Nov., fr. Feb.-Jun. of next year.

幼苗 Seedling

摄影：方腾 Photo by: Fang Teng

花序 Inflorescence

摄影：方腾 Photo by: Fang Teng

核果 Drupe

摄影：方腾 Photo by: Fang Teng

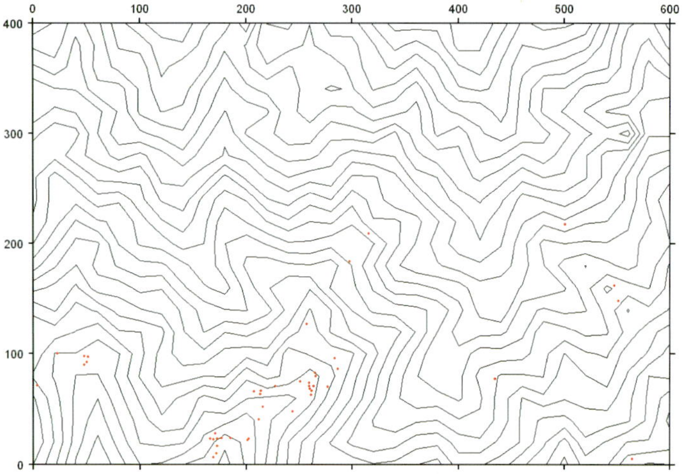

个体分布图 Distribution of Individuals

径级分布表 DBH Class

胸径区间 (Diameter class) (cm)	个体数 (No. of individuals in the plot)	比例 (Proportion)
1～2	19	0.43
2～5	15	0.34
5～10	7	0.16
10～20	3	0.07
20～30	0	0.00
30～60	0	0.00
≥60	0	0.00

33 黄绒润楠（黄桢楠） huángróngrùnnán(huángzhēnnán) | Grijs Machilus

Machilus grijsii Hance
樟科 | Lauraceae

代码（SpCode）= MACGRI
个体数（Individual number/24ha）= 449
最大胸径（Max DBH）= 7.8 cm
重要值排序（Importance value rank）= 57

常绿小乔木，高达5m。树皮灰色。芽、小枝、叶柄、叶背有黄褐色短绒毛。叶革质。花序短，束状生于枝顶。核果球形，嫩时绿色，熟时蓝黑色。花期3月，果期5月。

Evergreen small tree, up to 5 m tall. Buds, branchlets, petioles, and leaf blade abaxially yellow-brown velutinous. Leaf blade leathery. Inflorescence short, fasciculate at apex of branchlet. Fruit globose, green at young and dark blue at maturity. Fl. Mar., fr. May.

花枝 Flowering branch
摄影：陈建华 Photo by: Chen Jianhua

叶背 Leaf back
摄影：方腾 Photo by: Fang Teng

果枝 Fruiting branch
摄影：杨波 Photo by: Yang Bo

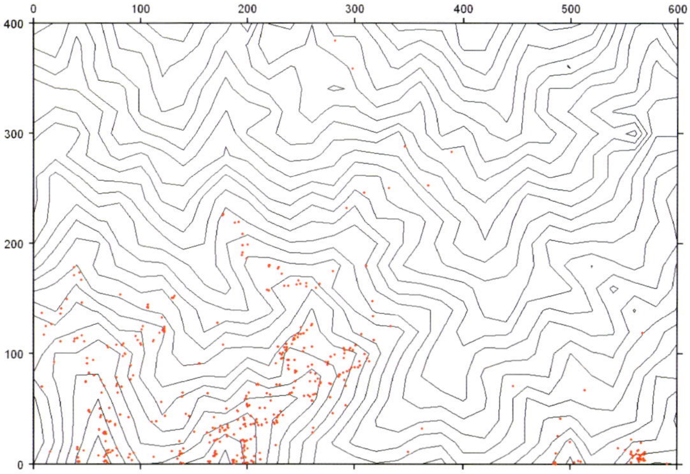

个体分布图 Distribution of Individuals

径级分布表 DBH Class

胸径区间 (Diameter class) (cm)	个体数 (No. of individuals in the plot)	比例 (Proportion)
1～2	372	0.83
2～5	76	0.17
5～10	1	0.00
10～20	0	0.00
20～30	0	0.00
30～60	0	0.00
≥60	0	0.00

34 薄叶润楠（华东楠） báoyèrùnnán(huádōngnán) | Thinleaf Machilus

Machilus leptophylla Hand.-Mazz.
樟科 | Lauraceae

代码（SpCode）= MACLEP
个体数（Individual number/24ha）= 1
最大胸径（Max DBH）= 1.7 cm
重要值排序（Importance value rank）= 155

常绿高大乔木，高达28m。树皮灰褐色。枝粗壮，暗褐色，无毛。叶薄纸质。花序集生在幼枝基部。核果球形，熟时紫黑色，果梗肉质，鲜红色。花期4月，果期7月。

Evergreen large tree, up to 28 m tall. Bark grayish brown. Branchlets thick, deeply brown, glabrous. Leaf blade thinly papery. Panicles many, congested on base of young branchlet. Fruit globose, dark purple; fruiting pedicel fleshy, red at maturity. Fl. Apr., fr. Jul..

幼苗 Seedling
摄影：方腾 Photo by: Fang Teng

叶 Leaf
摄影：陈彬 Photo by: Chen Bin

花序 Inflorescence
摄影：陈彬 Photo by: Chen Bin

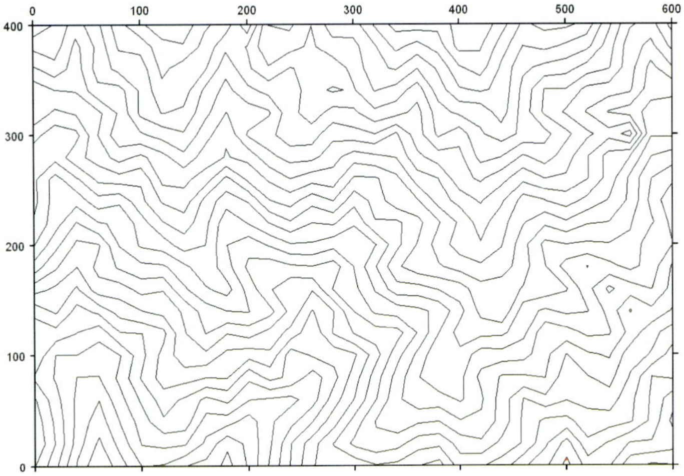

个体分布图 Distribution of Individuals

径级分布表 DBH Class

胸径区间 (Diameter class) (cm)	个体数 (No. of individuals in the plot)	比例 (Proportion)
1～2	1	1.00
2～5	0	0.00
5～10	0	0.00
10～20	0	0.00
20～30	0	0.00
30～60	0	0.00
≥60	0	0.00

35 刨花润楠 bàohuārùnnán | Baohua Machilus

Machilus pauhoi Kanehira
樟科 | Lauraceae

代码（SpCode）= MACPAU
个体数（Individual number/24ha）= 2
最大胸径（Max DBH）= 13 cm
重要值排序（Importance value rank）= 140

常绿乔木，高达20m。树皮灰褐色。小枝绿带褐色。叶常集生于小枝梢端。叶革质，叶脉12～17对。核果球形，熟时黑色。花期3月，果期6月。

Evergreen tree, up to 20 m tall. Bark grayish brown. Branchlets green with little brown. Leaves usually congested at apex of branchlet; leaf blade leathery, lateral veins 12-17 pairs. Fruit globose, black when mature. Fl. Mar., fr. Jun..

幼苗 Seedling
摄影：金冬梅 Photo by: Jin Dongmei

叶 Leaf
摄影：张金龙 Photo by: Zhang Jinlong

叶背 Leaf back
摄影：张金龙 Photo by: Zhang Jinlong

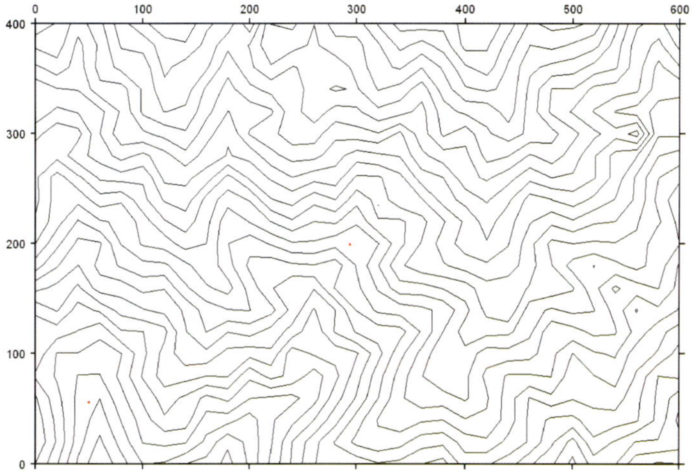

个体分布图 Distribution of Individuals

径级分布表 DBH Class

胸径区间 (Diameter class) (cm)	个体数 (No. of individuals in the plot)	比例 (Proportion)
1～2	0	0.00
2～5	0	0.00
5～10	1	0.50
10～20	1	0.50
20～30	0	0.00
30～60	0	0.00
≥60	0	0.00

36 红楠 hóngnán | Red Nanmu

Machilus thunbergii Sieb. et Zucc.
樟科 | Lauraceae

代码（SpCode）= MACTHU
个体数（Individual number/24ha）= 1384
最大胸径（Max DBH）= 43.2 cm
重要值排序（Importance value rank）= 22

常绿乔木，高达20m。树皮黄褐色。小枝绿色，一年生小枝无皮孔，二年生以上小枝疏生显著隆起皮孔。果扁球形，熟时紫黑色。果梗肉质增粗，鲜红色。花期3月，果期6~7月。

Evergreen tree, up to 20 m tall. Bark yellowish brown. Young branchlets purple-brown, without lenticel; older branchlets rough with lenticel. Leaf blade leathery. Fruit green becoming dark purple, compressed globose; peduncle and fruiting pedicel fleshy, reddish purple. Fl. Mar., fr. Jun.-Jul..

幼苗 Seedling
摄影：金冬梅 Photo by: Jin Dongmei

果枝 Fruiting branch
摄影：方腾 Photo by: Fang Teng

核果 Drupe
摄影：方腾 Photo by: Fang Teng

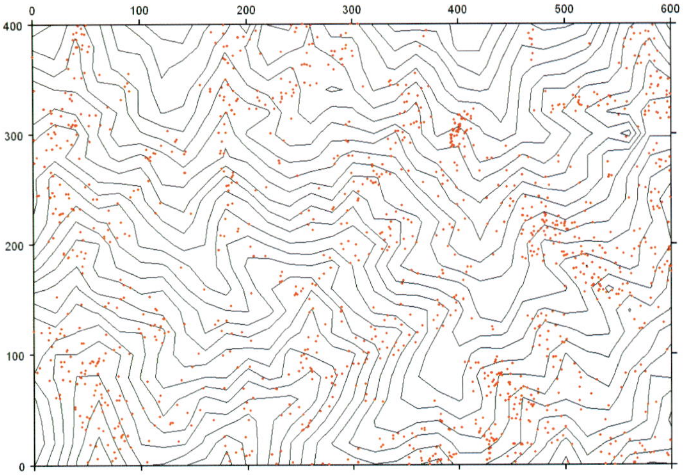

个体分布图 Distribution of Individuals

径级分布表 DBH Class

胸径区间 (Diameter class) (cm)	个体数 (No. of individuals in the plot)	比例 (Proportion)
1~2	494	0.36
2~5	355	0.26
5~10	109	0.08
10~20	187	0.14
20~30	190	0.14
30~60	49	0.04
≥60	0	0.00

37 浙江新木姜子 zhèjiāngxīnmùjiāngzǐ | Zhejiang Newlitsea

Neolitsea aurata (Hay.)Koidz. var. *chekiangensis* (Nakai) Y. C. Yang et P. H. Huang
樟科 | Lauraceae

代码（SpCode）= NEOAUR
个体数（Individual number/24ha）= 9098
最大胸径（Max DBH）= 40 cm
重要值排序（Importance value rank）= 5

常绿小乔木，高达14m。树皮灰色至深灰色，平滑。小枝灰绿色，被易脱落的锈褐色绢状毛。叶互生或者集生枝顶。核果熟时紫黑色，有光泽。花期3～4月，果期10～11月。

Evergreen tree, up to 14 m tall. Bark grey to dark grey, smooth. Young branchlets and petioles ferruginous or yellow-brown pubescent or glabrous. Leaves alternate or clustered toward apex of branchlet. Fruit ellipsoid, dark purple, lustrous. Fl. Mar.-Apr., fr. Oct.-Nov..

幼苗 Seedling
摄影：方腾 Photo by: Fang Teng

叶 Leaf
摄影：陈彬 Photo by: Chen Bin

幼果 Fruit(immature)
摄影：陈彬 Photo by: Chen Bin

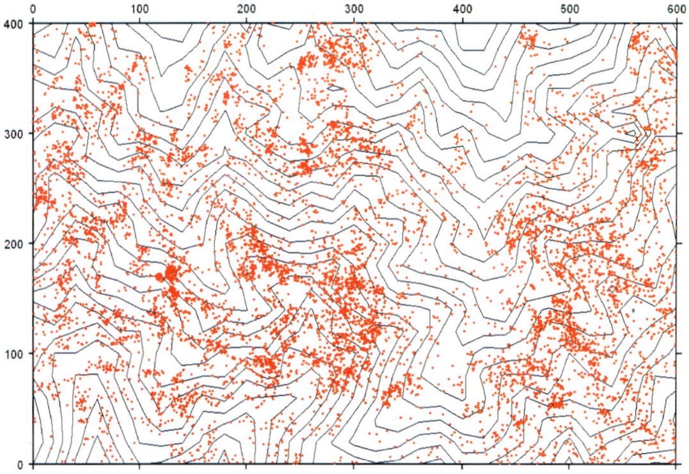

个体分布图 Distribution of Individuals

径级分布表 DBH Class

胸径区间 (Diameter class) (cm)	个体数 (No. of individuals in the plot)	比例 (Proportion)
1～2	4849	0.53
2～5	3592	0.39
5～10	598	0.07
10～20	58	0.01
20～30	0	0.00
30～60	1	0.00
≥60	0	0.00

38 檫木 chámù | Sassafras

Sassafras tzumu (Hemsl.) Hemsl.
樟科 | Lauraceae

代码（SpCode）= SASTZU
个体数（Individual number/24ha）= 4
最大胸径（Max DBH）= 43.5 cm
重要值排序（Importance value rank）= 129

落叶乔木，高达35m。树皮幼时黄绿色，老时灰褐色，不规则深纵裂。花黄色，先叶开放。核果近球形，熟时由红色变为蓝黑色，被白粉，生于红色浅杯状果托上。花期2～3月，果期7～8月。

Deciduous tree, up to 35 m tall. Bark yellow-green when young but gray-brown when mature, irregularly and longitudinally fissured. Flowers yellow, appear before leaves. Fruit subglobose, blue-black and white waxy when mature, seated on red shallow perianth cup. Fl. Feb.-Mar., fr. Jul.-Aug..

幼苗 Seedling
摄影：金冬梅 Photo by: Jin Dongmei

枝叶 Branch and leaves
摄影：杨波 Photo by: Yang Bo

核果 Drupe
摄影：杨波 Photo by: Yang Bo

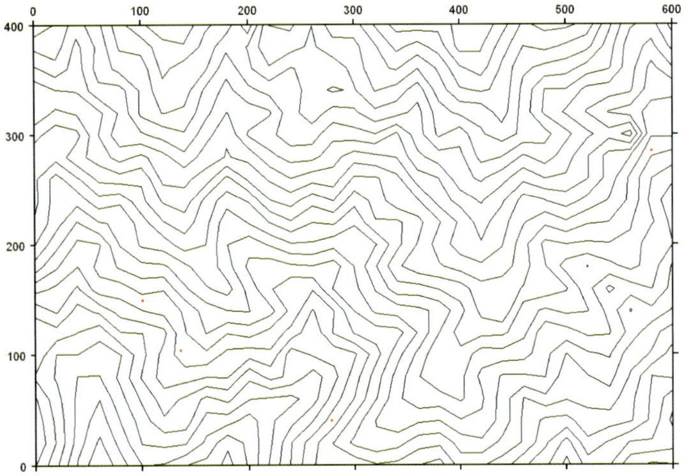

个体分布图 Distribution of Individuals

径级分布表 DBH Class

胸径区间 (Diameter class) (cm)	个体数 (No. of individuals in the plot)	比例 (Proportion)
1～2	0	0.00
2～5	0	0.00
5～10	0	0.00
10～20	1	0.25
20～30	1	0.25
30～60	2	0.50
≥60	0	0.00

39 中国绣球（伞八仙） zhōngguóxiùqiú(sǎnbāxiān) | China Hydrangea

Hydrangea chinensis Maxim.
虎耳草科 | Saxifragaceae

代码（SpCode）= HYDCHI
个体数（Individual number/24ha）= 58
最大胸径（Max DBH）= 2.4 cm
重要值排序（Importance value rank）= 72

半常绿灌木，高1～4m。小枝灰黄色至红褐色，被稀疏粗伏毛，后变无毛。叶纸质。伞形聚伞花序宽3～7cm，果期宽10～14cm。蒴果纺锤形到卵球形。花期3～8月，果期5～9月。

Semi-evergreen shrub 1-4 m tall. First and 2nd year's branchlets red-brown to brown, pubescent when young, glabrescent. Leaf blade papery. Inflorescences umbellate or subcorymbose, 3-7 cm, 10-14 cm wide in fruit. Capsule fusiform to ovoid-globose. Fl. Mar.-Aug., fr. May-Sep..

幼苗 Seedling
摄影：方腾 Photo by: Fang Teng

小枝 Branchlet
摄影：张金龙 Photo by: Zhang Jinlong

花序 Inflorescence
摄影：陈彬 Photo by: Chen Bin

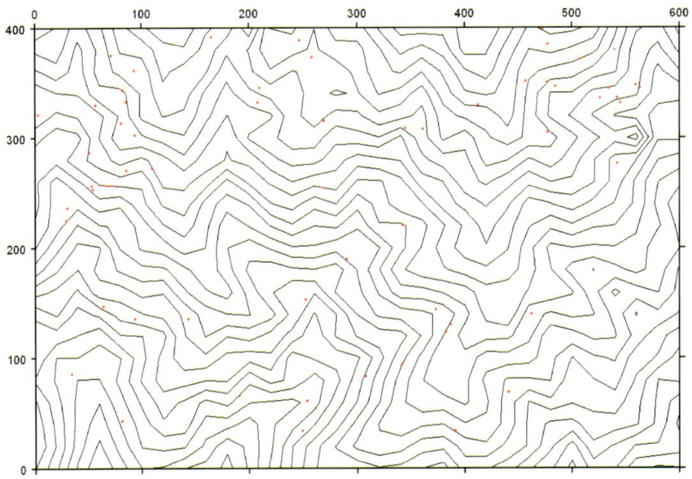

个体分布图 Distribution of Individuals

径级分布表 DBH Class

胸径区间 (Diameter class) (cm)	个体数 (No. of individuals in the plot)	比例 (Proportion)
1～2	56	0.97
2～5	2	0.03
5～10	0	0.00
10～20	0	0.00
20～30	0	0.00
30～60	0	0.00
≥60	0	0.00

40 峨眉鼠刺（长圆叶鼠刺） éméishǔcì(chángyuányèshǔcì) | Emei Sweetspire

Itea omeiensis Schneid.

虎耳草科 | Saxifragaceae

代码（SpCode）= ITEOME

个体数（Individual number/24ha）= 1334

最大胸径（Max DBH）= 7.5 cm

重要值排序（Importance value rank）= 27

常绿灌木或小乔木，高达10m。幼枝黄绿色，光滑，老枝棕褐色，有纵棱。叶厚革质，两面无毛，侧脉5～7对。蒴果长6～9mm。花期3～5月，果期6～11月。

Evergreen shrub or small tree, up to 10 m tall. Young branchlets yellow-green, glabrous; old branchlets brown, striate. Leaf blade thickly leathery, both surfaces glabrous, secondary veins in 5-7 pairs. Capsule 6-9 mm, pubescent. Fl. Mar.-May, fr. Jun.-Nov..

幼苗 Seedling

摄影：金冬梅 Photo by: Jin Dongmei

叶 Leaf

摄影：张金龙 Photo by: Zhang Jinlong

花序 Inflorescence

摄影：方腾 Photo by: Fang Teng

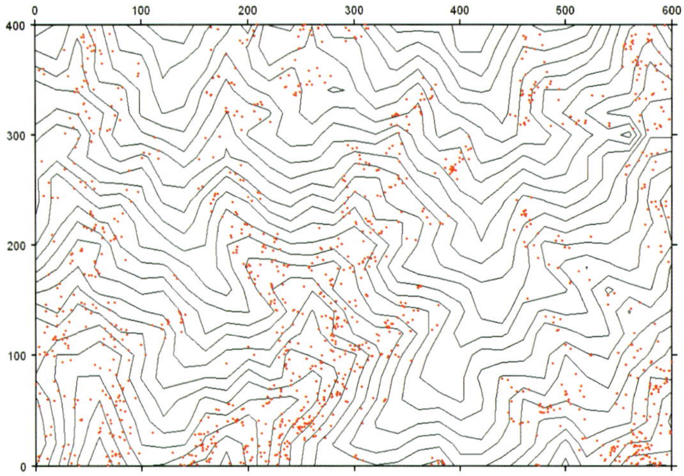

个体分布图 Distribution of Individuals

径级分布表 DBH Class

胸径区间 (Diameter class) (cm)	个体数 (No. of individuals in the plot)	比例 (Proportion)
1～2	883	0.66
2～5	429	0.32
5～10	22	0.02
10～20	0	0.00
20～30	0	0.00
30～60	0	0.00
≥60	0	0.00

41 海金子　hǎijīnzi | Anisetree-like Seatung

Pittosporum illicioides Makino
海桐花科 | Pittosporaceae

代码（SpCode）= PITILL
个体数（Individual number/24ha）= 52
最大胸径（Max DBH）= 2.5 cm
重要值排序（Importance value rank）= 76

常绿灌木，高达 5m。嫩枝光滑，老枝有皮孔。叶生于枝顶，簇生呈假轮生状。伞形花序顶生，花黄色。蒴果近圆形，3 片裂开，种子外面包覆红色粘质假种皮。花期 3 ~ 5 月，果期 7 ~ 9 月。

Evergreen shrub, up to 5 m tall. Young branchlets glabrous; old branchlets lenticellate. Leaves clustered at branchlet apex, appearing pseudoverticillate. Inflorescences terminal, flowers yellow. Capsule subglobose. Seeds covered by tachy red aril. Fl. Mar.-May, fr. Jul.-Sep..

幼苗　Seedling
摄影：金冬梅　Photo by: Jin Dongmei

花　Flower
摄影：方腾　Photo by: Fang Teng

蒴果　Capsule
摄影：陈国科　Photo by: Chen Guoke

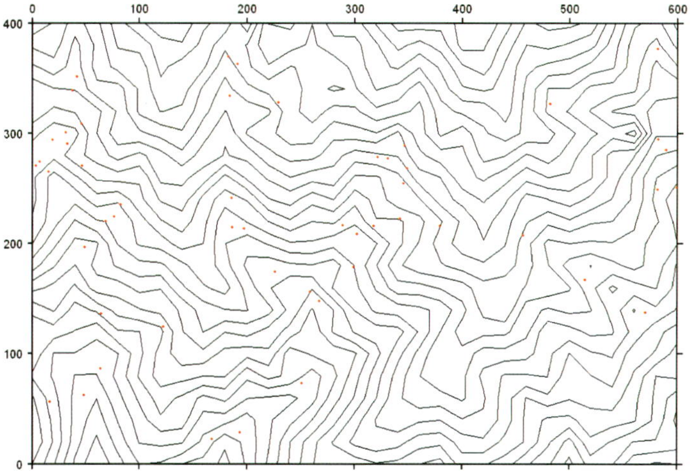

个体分布图 Distribution of Individuals

径级分布表 DBH Class

胸径区间 (Diameter class) (cm)	个体数 (No. of individuals in the plot)	比例 (Proportion)
1～2	49	0.94
2～5	3	0.06
5～10	0	0.00
10～20	0	0.00
20～30	0	0.00
30～60	0	0.00
≥60	0	0.00

42 腺蜡瓣花（灰白蜡瓣花） xiànlàbànhuā(huībáilàbànhuā) | Glaucousback Waxpetal

Corylopsis glandulifera Hemsl.
金缕梅科 | Hamamelidaceae

代码（SpCode）= CORGLA
个体数（Individual number/24ha）= 3343
最大胸径（Max DBH）= 10.6 cm
重要值排序（Importance value rank）= 34

落叶灌木，高达3m。嫩枝光滑，老枝有皮孔。叶下面有时灰白色，无毛。总状花序顶生，花先叶开放。蒴果光滑，种子3～4mm，黑色。花期2～3月，果期6～8月。

Deciduous shrub, up to 3 m tall; young branches glabrous, older lenticellate. Leaf blade abaxially gray-white, glabrous. Raceme terminal on shoots, appear before leaves. Capsules glabrous. Seeds 3-4 mm, black. Fl. Feb.-Mar., fr. Jun.-Aug..

叶 Leaf
摄影：方腾 Photo by: Fang Teng

花序 Inflorescence
摄影：陈建华 Photo by: Chen Jianhua

蒴果 Capsule
摄影：陈建华 Photo by: Chen Jianhua

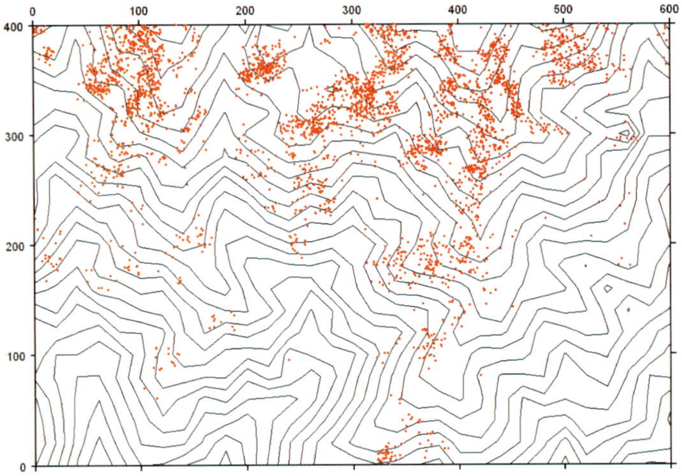

个体分布图 Distribution of Individuals

径级分布表 DBH Class

胸径区间 (Diameter class) (cm)	个体数 (No. of individuals in the plot)	比例 (Proportion)
1～2	495	0.15
2～5	2274	0.68
5～10	572	0.17
10～20	1	0.00
20～30	0	0.00
30～60	1	0.00
≥60	0	0.00

43 杨梅叶蚊母树 yángméiyèwénmŭshù | Myrica-like Mosquitomam

Distylium myricoides Hemsl.
金缕梅科 | Hamamelidaceae

代码（SpCode）= DISMYR
个体数（Individual number/24ha）= 3466
最大胸径（Max DBH）= 25 cm
重要值排序（Importance value rank）= 36

常绿灌木或小乔木，嫩枝有鳞垢，老枝灰褐色，无毛而有皮孔。花序长1～3cm，花药红色，长2～3cm。蒴果卵球或卵形，被黄棕色星状毛。花期4～6月，果期9～10月。

Evergreen shrub or small tree; young branches stellately lepidote, older growth drying gray-brown, glabrescent, lenticellate; leaf blade both surfaces glabrous, leathery. Inflorescences 1-3 cm; anthers red. Capsules ovoid-globose or subovoid, stellately pubescent with yellow-brown hairs. Fl. Apr.-Jun., fr. Sep.-Oct..

幼苗 Seedling
摄影：金冬梅 Photo by: Jin Dongmei

叶与虫瘿 Leaf and insect gall
摄影：方腾 Photo by: Fang Teng

蒴果 Capsule
摄影：方腾 Photo by: Fang Teng

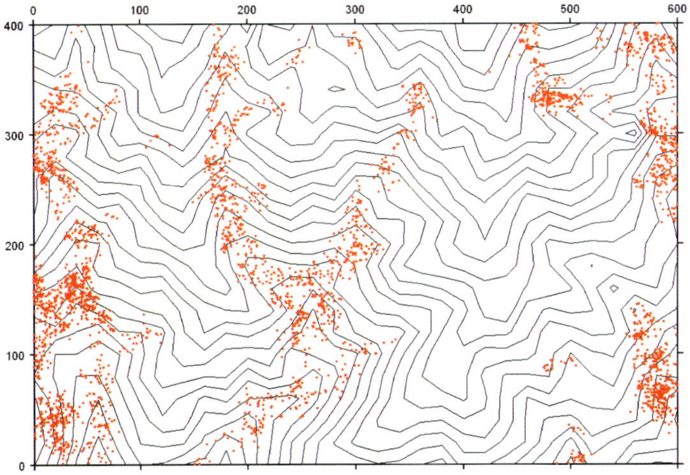

个体分布图 Distribution of Individuals

径级分布表 DBH Class

胸径区间 (Diameter class) (cm)	个体数 (No. of individuals in the plot)	比例 (Proportion)
1～2	1450	0.42
2～5	885	0.26
5～10	650	0.19
10～20	476	0.14
20～30	7	0.00
30～60	0	0.00
≥60	0	0.00

44 枫香树

fēngxiāngshù | Beautiful Sweetgum

Liquidambar formosana Hance
金缕梅科 | Hamamelidaceae

代码（SpCode）= LIQFOR
个体数（Individual number/24ha）= 38
最大胸径（Max DBH）= 29.2 cm
重要值排序（Importance value rank）= 79

落叶大乔木，高达30m。树皮灰褐色，方块状剥落。小枝略有皮孔。叶革质或厚纸质。头状花序。雄性花序常多个排成总状。果序圆球形，蒴果有宿存刺状退化雄蕊。花期3月，果期8～10月。

Deciduous big tree, up to 30 m tall. Bark gray-brown, squarely exfoliate; branchlets somewhat lenticellate; leaf blade leathery or thick papery. Male inflorescence several arranged in a raceme. Infructescence globose. Capsules with persistent spinous staminodes. Fl. Mar., fr. Aug.-Oct..

花序 Inflorescence
摄影：杨波 Photo by: Yang Bo

叶 Leaf
摄影：金冬梅 Photo by: Jin Dongmei

果序 Infructescence
摄影：方腾 Photo by: Fang Teng

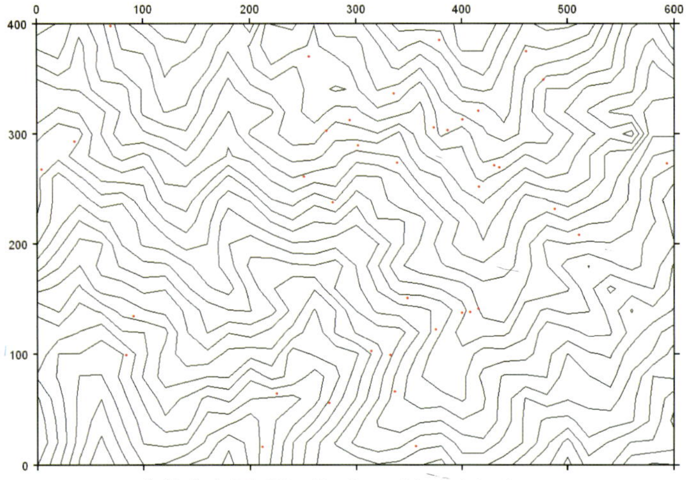

个体分布图 Distribution of Individuals

径级分布表 DBH Class

胸径区间 (Diameter class) (cm)	个体数 (No. of individuals in the plot)	比例 (Proportion)
1～2	0	0.00
2～5	0	0.00
5～10	3	0.08
10～20	26	0.68
20～30	9	0.24
30～60	0	0.00
≥60	0	0.00

45 檵木 jìmù | China Loropetal

Loropetalum chinense (R. Br.) Oliv.
金缕梅科 | Hamamelidaceae

代码（SpCode）= LORCHI
个体数（Individual number/24ha）= 4461
最大胸径（Max DBH）= 24.7 cm
重要值排序（Importance value rank）= 10

常绿灌木，有时为小乔木，多分枝，小枝有星毛。花序短总状或头状；花先叶开放，花瓣4或6，白色，淡黄色或红色。子房下位。蒴果卵形或卵球形。花期3～4月，果期6～10月。

Evergreen shrub or small tree, much branched; branchlets stellately pubescent. Inflorescence a short raceme or nearly capitate. Flowers open before leaves appear. Petals 4 or 6, white, pale yellow or red. Ovary inferior. Capsules ovoid or obovoid-globose. Fl. Mar.-Apr., fr. Jun.-Oct..

幼苗 Seedling
摄影：方腾 Photo by: Fang Teng

花 Flower
摄影：方腾 Photo by: Fang Teng

蒴果 Capsule
摄影：张金龙 Photo by: Zhang Jinlong

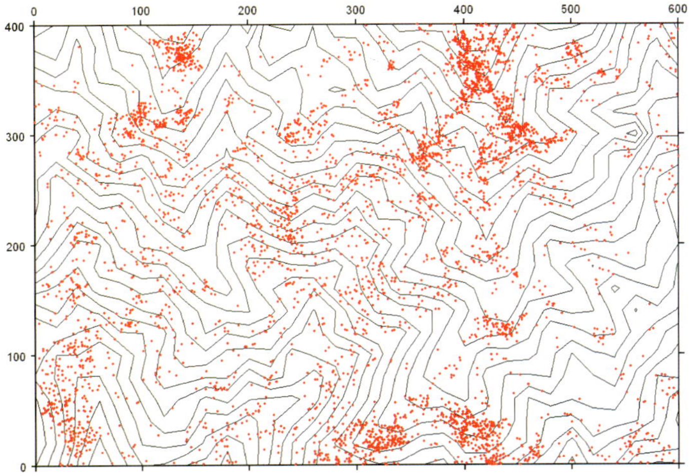

个体分布图 Distribution of Individuals

径级分布表 DBH Class

胸径区间 (Diameter class) (cm)	个体数 (No. of individuals in the plot)	比例 (Proportion)
1～2	1307	0.29
2～5	1649	0.37
5～10	846	0.19
10～20	645	0.14
20～30	13	0.00
30～60	0	0.00
≥60	1	0.00

46 光萼林檎（尖嘴林檎） guāngèlínqín(jiānzuǐlínqín) | Smooth Calyx Apple

Malus leiocalyca S. Z. Huang
蔷薇科 | Rosaceae

代码（SpCode）= MALLEI
个体数（Individual number/24ha）= 161
最大胸径（Max DBH）= 24 cm
重要值排序（Importance value rank）= 56

落叶灌木或小乔木，高4～10m。老枝深灰色，微弯曲。叶嫩时微被毛，老时光滑。花序伞形，5～7花。梨果绿色，熟时黄红色，直径1.5～2.5cm，萼宿存。花期5月，果期8～9月。

Deciduous shrub or small tree, 4-10 m tall. Branchlets dark gray when old, slightly curved. Leaf blade sparsely pubescent when young, glabrescen. Corymb umbel-like, 5-7-flowered. Pome green when young, yellowish red at maturity, globose, 1.5-2.5 cm in diam; sepals persistent. Fl. May, fr. Aug.-Sep..

树干 Trunk
摄影：方腾 Photo by: Fang Teng

枝叶 Branch and leaves
摄影：陈建华 Photo by: Chen Jianhua

梨果 Pome
摄影：方腾 Photo by: Fang Teng

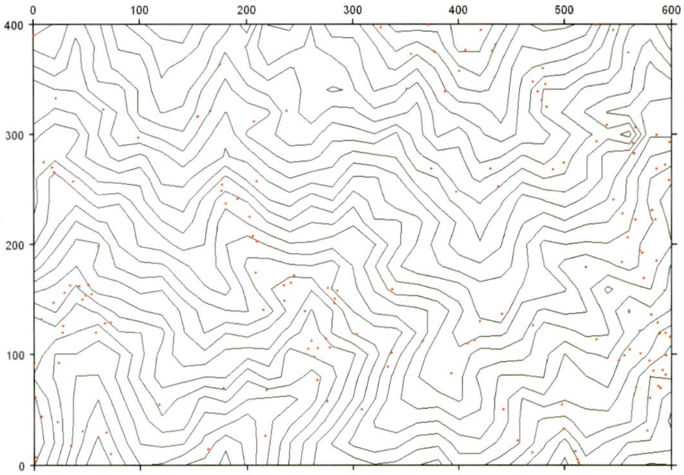

个体分布图 Distribution of Individuals

径级分布表 DBH Class

胸径区间 (Diameter class) (cm)	个体数 (No. of individuals in the plot)	比例 (Proportion)
1～2	15	0.09
2～5	12	0.07
5～10	46	0.29
10～20	81	0.50
20～30	7	0.04
30～60	0	0.00
≥60	0	0.00

47 中华石楠 zhōnghuáshínán | China Photinia

Photinia beauverdiana Schneid.
蔷薇科 | Rosaceae

代码（SpCode）= PHOBEA
个体数（Individual number/24ha）= 34
最大胸径（Max DBH）= 7.8 cm
重要值排序（Importance value rank）= 88

落叶灌木或小乔木，高 3 ~ 10m。小枝紫褐色到黑褐色，有散生灰色皮孔。叶片薄纸质。复伞房花序顶生，总花梗和花梗无毛，密生长形皮孔。果实紫红色，卵形或近球形。花期 5 月，果期 7 ~ 8 月。

Deciduous shrub or small tree, 3-10 m tall. Branchlets purplish brown to blackish brown, with scattered gray lenticels; leaf blade thin papery. Compound corymbs terminal, rachis and pedicels glabrous, with numerous oblong lenticels. Fruit purplish red, ovoid or subglobose. Fl. May, fr. Jul.-Aug..

幼苗 Seedling
摄影：方腾 Photo by: Fang Teng

花序 Flower
摄影：陈彬 Photo by: Chen Bin

梨果 Pome
摄影：陈彬 Photo by: Chen Bin

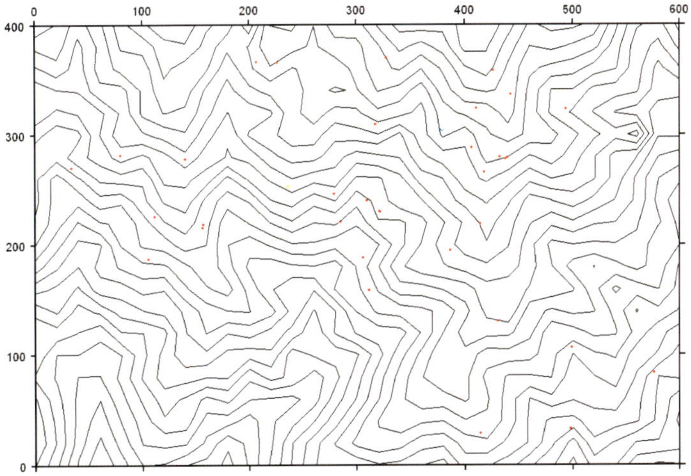

个体分布图 Distribution of Individuals

径级分布表 DBH Class

胸径区间 (Diameter class) (cm)	个体数 (No. of individuals in the plot)	比例 (Proportion)
1～2	13	0.38
2～5	13	0.38
5～10	8	0.24
10～20	0	0.00
20～30	0	0.00
30～60	0	0.00
≥60	0	0.00

48 光叶石楠 guāngyèshínán | Smoothleaf Photinia

Photinia glabra (Thunb.) Maxim.
蔷薇科 | Rosaceae

代码（SpCode）= PHOGLA
个体数（Individual number/24ha）= 780
最大胸径（Max DBH）= 19 cm
重要值排序（Importance value rank）= 26

常绿小乔木，高3～5m。嫩枝褐色到灰褐色，老枝灰黑色。叶片革质，嫩叶及老叶皆呈红色。花多数，成顶生复伞房花序。果实红色，倒卵形到卵形。花期4～5月，果期9～10月。

Evergreen tree, 3-5 m tall. Branchlets brown to grayish brown when young, grayish black when old. Leaf blade leathery, initially and oldly reddish. Compound corymbs terminal, numerous flowered. Fruit red, obovate or ovoid. Fl. Apr.-May, fr. Sep.-Oct..

果枝 Fruiting branch
摄影：张金龙 Photo by: Zhang Jinlong

幼苗 Seedling
摄影：金冬梅 Photo by: Jin Dongmei

叶 Leaf
摄影：张金龙 Photo by: Zhang Jinlong

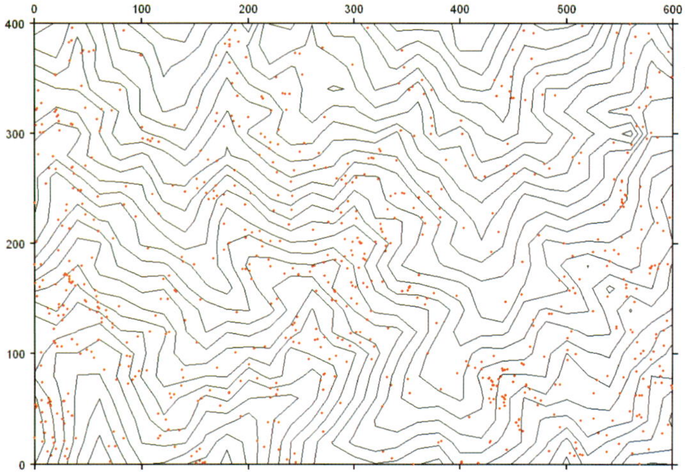

个体分布图 Distribution of Individuals

径级分布表 DBH Class

胸径区间 (Diameter class) (cm)	个体数 (No. of individuals in the plot)	比例 (Proportion)
1～2	367	0.47
2～5	290	0.37
5～10	111	0.14
10～20	14	0.02
20～30	0	0.00
30～60	0	0.00
≥60	0	0.00

49 小叶石楠 xiǎoyèshínán | Littleleaf Photinia

Photinia parvifolia (E. Pritz.) Schneid.
蔷薇科 | Rosaceae

代码（SpCode）= PHOPAR
个体数（Individual number/24ha）= 39
最大胸径（Max DBH）= 8.6 cm
重要值排序（Importance value rank）= 89

落叶灌木，高1～3m。枝纤细，小枝红褐色，老枝黑褐色。叶片纸质。花数朵到略多于10朵，生于侧枝顶端。果实橘红色或紫色，有宿存直立萼片。果柄长1～2.5cm，密生皮孔。花期4～5月，果期7～8月。

Deciduous shrub, 1-3 m tall. Branchlets reddish brown, blackish blown when old, slender; leaf blade papery. Inflorescences terminating lateral branchlets, flowers several to slightly more than 10. Fruit orangish red or purple, sepals erect; fruiting pedicels 1-2.5 cm, with dense lenticels. Fl. Apr.-May, fr. Jul.-Aug..

花枝 Flowering branch
摄影：陈彬 Photo by: Chen Bin

花序 Inflorescence
摄影：方腾 Photo by: Fang Teng

果枝 Fruiting branch
摄影：严岳鸿 Photo by: Yan Yuehong

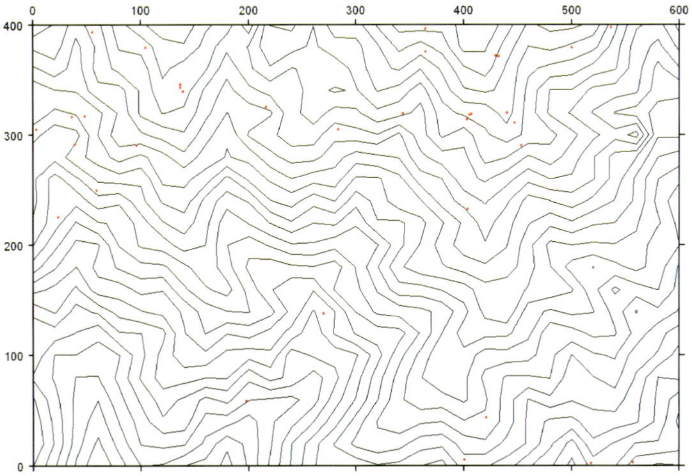

个体分布图 Distribution of Individuals

径级分布表 DBH Class

胸径区间 (Diameter class) (cm)	个体数 (No. of individuals in the plot)	比例 (Proportion)
1～2	19	0.49
2～5	14	0.36
5～10	6	0.15
10～20	0	0.00
20～30	0	0.00
30～60	0	0.00
≥60	0	0.00

50 石楠 shínán | Photinia

Photinia serratifolia (Desf.) Kalkman
蔷薇科 | Rosaceae

代码（SpCode）= PHOSER
个体数（Individual number/24ha）= 18
最大胸径（Max DBH）= 28 cm
重要值排序（Importance value rank）= 104

常绿灌木或小乔木，高4～6m。枝嫩时红褐色，老时褐灰色。叶片革质，侧脉20～30对。复伞房花序顶生，花密生。果实球形，红色，后呈紫褐色，种子1枚。花期4～5月，果期9～10月。

Evergreen shrub or small tree, 4-6 m tall. Branchlets brown or reddish brown when young, brownish gray when old. Leaf blade leathery, veins 20-30 pairs. Compound corymbs terminal, many flowers clustered. Fruit red when immature, brownish purple when mature, globose, with 1 seed. Fl. Apr.-May, fr. Sep.-Oct..

嫩叶 New leaves
摄影：陈彬 Photo by: Chen Bin

叶 Leaf
摄影：张金龙 Photo by: Zhang Jinlong

花序 Inflorescence
摄影：陈彬 Photo by: Chen Bin

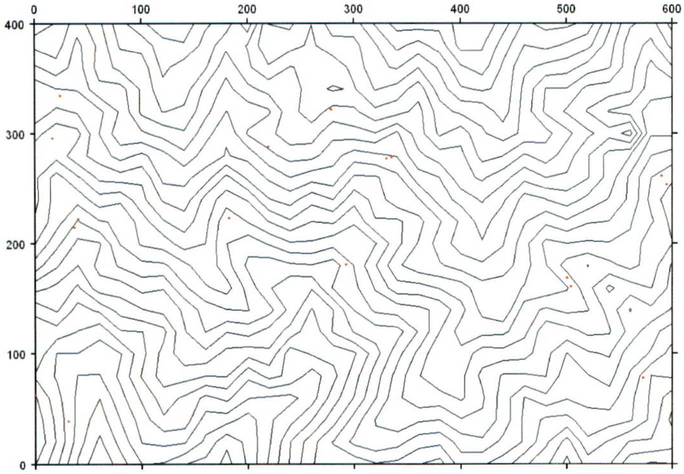

个体分布图 Distribution of Individuals

径级分布表 DBH Class

胸径区间 (Diameter class) (cm)	个体数 (No. of individuals in the plot)	比例 (Proportion)
1～2	6	0.35
2～5	1	0.06
5～10	2	0.12
10～20	7	0.41
20～30	1	0.06
30～60	0	0.00
≥60	0	0.00

51 浙闽樱桃

zhèmǐnyīngtáo | Schneider Cherry

Cerasus schneideriana (Koehne) T. T. Yu & C. L. Li
蔷薇科 | Rosaceae

代码（SpCode）= CERSCH
个体数（Individual number/24ha）= 46
最大胸径（Max DBH）= 18.5 cm
重要值排序（Importance value rank）= 75

落叶小乔木，高2.5～6m。小枝紫褐色，嫩枝灰绿色。叶柄密被灰褐色微硬毛，先端有2～3枚黑色腺体。花序伞形，花1～3朵。核果紫红色，椭圆形。花期3月，果期5月。

Deciduous small tree, 2.5-6 m tall. Branchlets purplish brown; young branchlets grayish green. Petiole densely brown hirtellous, apex with 2 or 3 black nectaries. Inflorescences umbellate, 1-3-flowered. Drupe purplish red, ellipsoid. Fl. Mar., fr. May.

幼苗 Seedling
摄影：金冬梅 Photo by: Jin Dongmei

果枝 Fruiting branch
摄影：陈彬 Photo by: Chen Bin

核果 Drupe
摄影：陈彬 Photo by: Chen Bin

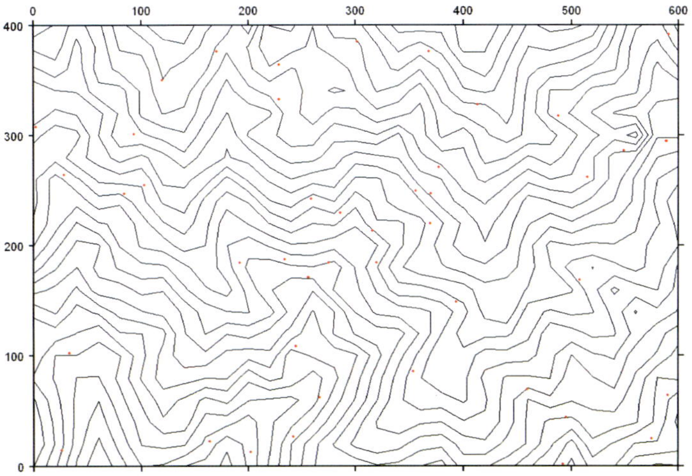

个体分布图 Distribution of Individuals

径级分布表 DBH Class

胸径区间 (Diameter class) (cm)	个体数 (No. of individuals in the plot)	比例 (Proportion)
1～2	4	0.09
2～5	9	0.20
5～10	17	0.37
10～20	15	0.33
20～30	0	0.00
30～60	1	0.02
≥60	0	0.00

52 刺叶桂樱

cìyèguìyīng | Spinyleaf Cherrylaurel

Laurocerasus spinulosa (Sieb. et Zucc.) Schneid.
蔷薇科 | Rosaceae

代码（SpCode）= LAUSPI
个体数（Individual number/24ha）= 43
最大胸径（Max DBH）= 25.3 cm
重要值排序（Importance value rank）= 83

常绿灌木或乔木，高达20m，偶尔呈灌木状。小枝紫褐色或黑褐色，具皮孔。叶片草质至薄革质，两面光滑。总状花序单生于叶腋。核果椭圆形，熟时褐色至黑褐色。花期9～10月，果期11月到翌年3月。

Evergreen tree, up to 20 m tall, rarely shrub. Branchlets purplish brown to blackish brown, with lenticels. Leaf herbaceous to thinly leathery, both surfaces glabrous. Racemes axillary, solitary. Drupe brown to blackish brown, ellipsoid. Fl. Sep.-Oct., fr. Nov. to Mar. of next year.

小枝 Branchlet
摄影：宋凯 Photo by: Song Kai

幼苗 Seedling
摄影：金冬梅 Photo by: Jin Dongmei

花序 Inflorescence
摄影：张代贵 Photo by: Zhang Daigui

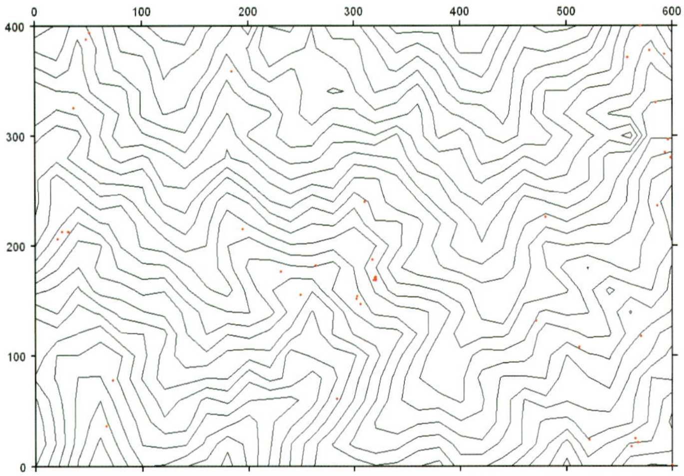

个体分布图 Distribution of Individuals

径级分布表 DBH Class

胸径区间 (Diameter class) (cm)	个体数 (No. of individuals in the plot)	比例 (Proportion)
1～2	11	0.26
2～5	13	0.30
5～10	7	0.16
10～20	9	0.21
20～30	3	0.07
30～60	0	0.00
≥60	0	0.00

53 石斑木

shíbānmù | Hongkong Hawthorn; India Hawthorn; Hongkong Raphiolepis

Rhaphiolepis indica (L.) Lindl.
蔷薇科 | Rosaceae

代码（SpCode）= RHAIND
个体数（Individual number/24ha）= 1993
最大胸径（Max DBH）= 6.2 cm
重要值排序（Importance value rank）= 17

常绿灌木。幼枝初被褐色绒毛，后逐渐脱落。叶革质，集生于枝顶，背面网脉明显。顶生圆锥花序或总状花序。梨果球形，熟时紫黑色，具花萼凋落的环状痕迹。花期4月，果期7～8月。

Evergreen shrub, up to 4 m tall. Branchlets initially brown tomentose, glabrous when old. Leaf clustered at branchlet apex, leathery, veins prominent abaxially. Panicle or racemes terminal. Pome purplish black, globose, sepals caducous, leaving an annular ring. Fl. Apr., fr. Jul.-Aug..

幼苗 Seedling
摄影：金冬梅 Photo by: Jin Dongmei

花序 Inflorescence
摄影：杨波 Photo by: Yang Bo

果枝 Fruiting branch
摄影：张金龙 Photo by: Zhang Jinlong

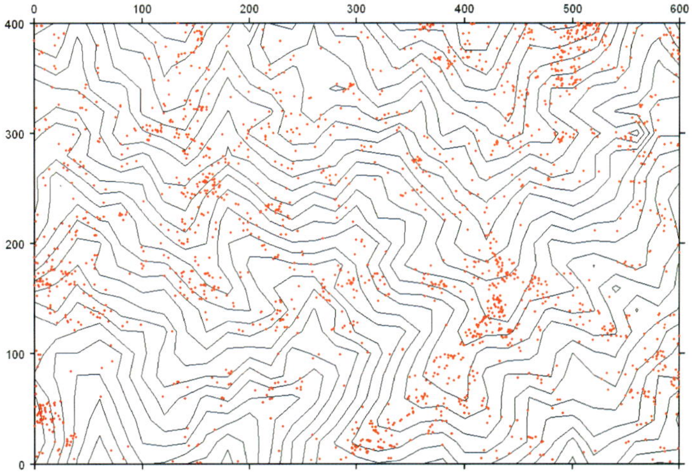

个体分布图 Distribution of Individuals

径级分布表 DBH Class

胸径区间 (Diameter class) (cm)	个体数 (No. of individuals in the plot)	比例 (Proportion)
1～2	1589	0.80
2～5	397	0.20
5～10	9	0.00
10～20	0	0.00
20～30	0	0.00
30～60	0	0.00
≥60	0	0.00

54 掌叶覆盆子 zhǎngyèfùpénzi | Palmleaf Raspberry

Rubus chingii Hu

蔷薇科 | Rosaceae

代码（SpCode）= RUBCHI

个体数（Individual number/24ha）= 2

最大胸径（Max DBH）= 1.4 cm

重要值排序（Importance value rank）= 152

落叶灌木，高1.5～3m，地上部分结果后枯死。嫩枝绿色，老时红褐色。具皮刺。单叶，掌状脉3～7条。花序单生于短侧枝顶，具1花。聚合果红色，直径1.5～2cm，密被灰色软毛。花期3～4月，果期5～6月。

Deciduous shrub, 1.5-3 m tall. Above-ground parts die after fruiting. Branchlets green when young, reddish brown in age, with sparse prickles. Leaves simple, palmately 3-7-veined. Inflorescences terminal on short branchlets, 1-flowered. Aggregate fruit red, subglobose, 1.5-2 cm in diam, densely gray pubescent. Fl. Mar.-Apr., fr. May-Jun..

幼苗 Seedling

摄影：金冬梅 Photo by: Jin Dongmei

花枝 Flowering branch

摄影：杨波 Photo by: Yang Bo

幼果 Fruit(immature)

摄影：陈彬 Photo by: Chen Bin

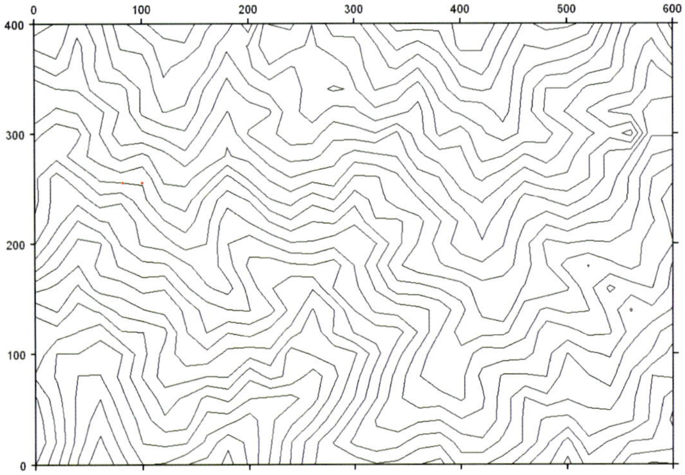

个体分布图 Distribution of Individuals

径级分布表 DBH Class

胸径区间 (Diameter class) (cm)	个体数 (No. of individuals in the plot)	比例 (Proportion)
1～2	2	1.00
2～5	0	0.00
5～10	0	0.00
10～20	0	0.00
20～30	0	0.00
30～60	0	0.00
≥60	0	0.00

55 石灰花楸

shíhuīhuāqiū | Lime Mountainash

Sorbus folgneri (Schneid.) Rehd.
蔷薇科 | Rosaceae

代码（SpCode）= SORFOL
个体数（Individual number/24ha）= 420
最大胸径（Max DBH）= 32.6 cm
重要值排序（Importance value rank）= 51

落叶乔木，高达10m。小枝圆柱形，黑褐色，幼时被白色绒毛，具少数椭圆形皮孔。单叶，上面深绿色，下面密被灰白色绒毛。复伞花序顶生。果实椭圆形，熟时红色或黄色。花期4～5月，果期7～8月。

Deciduous tree, up to 10 m tall. Branchlets blackish brown when old, terete, initially covered with grayish white tomentose, glabrescent, with few elliptic lenticels. Leaves simple, blade dark green adaxially, abaxially densely grayish white tomentose. Compound corymbs terminal. Fruit red or yellow, oblong or obovoid-oblong. Fl. Apr.-May, fr. Jul.-Aug..

叶背 Leaf back
摄影：金冬梅 Photo by: Jin Dongmei

幼苗 Seedling
摄影：邴艳红 Photo by: Bing Yanhong

果 Fruit
摄影：喻勋林 Photo by: Yu Xunlin

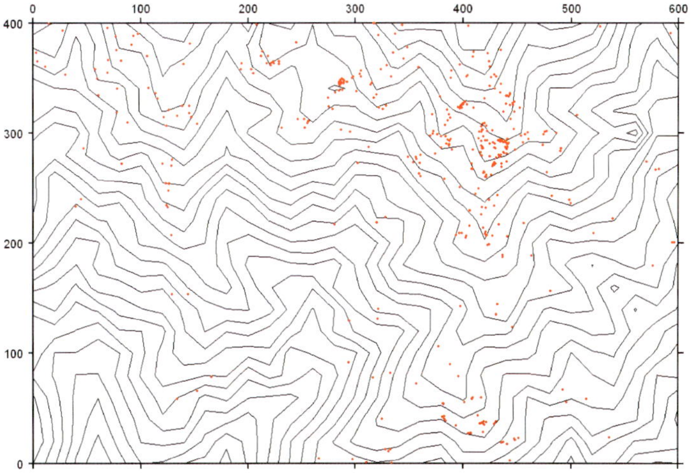

个体分布图 Distribution of Individuals

径级分布表 DBH Class

胸径区间 (Diameter class) (cm)	个体数 (No. of individuals in the plot)	比例 (Proportion)
1～2	175	0.42
2～5	149	0.35
5～10	32	0.08
10～20	37	0.09
20～30	25	0.06
30～60	2	0.00
≥60	0	0.00

56 山合欢（山槐） shānhéhuān (shānhuái) | Wild Siris

Albizia kalkora (Roxb.) Prain
豆科 | Leguminosae

代码（SpCode）= ALBKAL
个体数（Individual number/24ha）= 454
最大胸径（Max DBH）= 30.5 cm
重要值排序（Importance value rank）= 44

落叶乔木，高达20m。枝条暗褐色，有显著皮孔。二回羽状复叶。头状花序腋生或集生枝顶成圆锥花序。花初白色，后变黄或粉红色。荚果带状。花期5~6月，果期8~10月。

Deciduous tree, up to 20 m tall. Branchlets dark brown, with conspicuous lenticels. Leaves bipinnate. Capitulum axillary or clustered at branchlet apex into panicles. Flowers initially white then turn yellow or pink. Legume linearly. Fl. May-Jun., fr. Aug.-Oct..

荚果 Legume
摄影：方腾 Photo by: Fang Teng

幼苗 Seedling
摄影：金冬梅 Photo by: Jin Dongmei

花序 Inflorescence
摄影：方腾 Photo by: Fang Teng

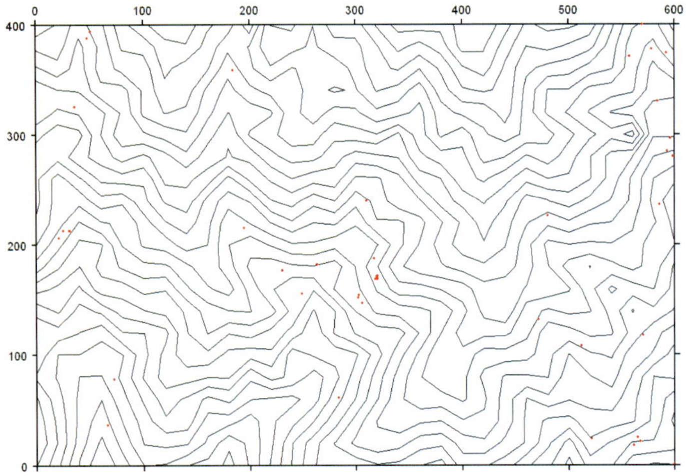

个体分布图 Distribution of Individuals

径级分布表 DBH Class

胸径区间 (Diameter class) (cm)	个体数 (No. of individuals in the plot)	比例 (Proportion)
1~2	14	0.03
2~5	113	0.25
5~10	195	0.43
10~20	107	0.24
20~30	23	0.05
30~60	2	0.00
≥60	0	0.00

57 黄檀 huángtán | Hubei Rosewood

Dalbergia hupeana Hance
豆科 | Leguminosae

代码（SpCode）= DALHUP
个体数（Individual number/24ha）= 261
最大胸径（Max DBH）= 32 cm
重要值排序（Importance value rank）= 48

落叶乔木，高10～20m。树皮暗灰色，呈薄片状剥落。羽状复叶，小叶3～5对，近革质，先端钝或凹入。花白色或淡紫色，荚果不裂，1～3粒种子。花期5～7月，果期8～10月。

Deciduous tree, 10-20 m tall. Bark dark-grey, scaly peel off. Leaves pinnately compound, leaflets 3-5 pairs, blade approximatively leathery, obtuse or concave at apex. Flowers white or mauve. Legume indehiscent with 1-3 seeds. Fl. May-Jul., fr. Aug.-Oct..

树干 Trunk
摄影：方腾 Photo by: Fang Teng

复叶 Compound Leaf
摄影：陈建华 Photo by: Chen Jianhua

荚果 Legume
摄影：陈彬 Photo by: Chen Bin

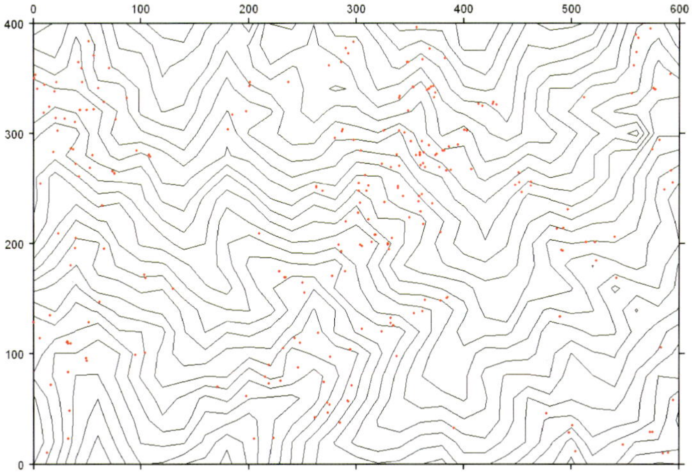

个体分布图 Distribution of Individuals

径级分布表 DBH Class

胸径区间 (Diameter class) (cm)	个体数 (No. of individuals in the plot)	比例 (Proportion)
1～2	42	0.16
2～5	48	0.18
5～10	57	0.22
10～20	98	0.37
20～30	16	0.06
30～60	1	0.00
≥60	0	0.00

58 美丽胡枝子 měilìhúzhīzǐ | Spiffy Bushclover

Lespedeza formosa (Vogel) Koehne
豆科 | Leguminosae

代码（SpCode）= LESFOR
个体数（Individual number/24ha）= 2
最大胸径（Max DBH）= 1.5 cm
重要值排序（Importance value rank）= 144

落叶灌木，高1～2m。分枝多，伸展，被疏柔毛。总状花序腋生，花冠紫色。荚果有网纹。花期7～9月，果期9～10月。

Deciduous shrub, 1-2 m tall; much branched, outstretched, with sparse pubescence. Raceme axillary, corolla purple. Legume with reticulated mottles. Fl. Jul.-Sep., fr. Sep.-Oct..

幼苗 Seedling
摄影：金冬梅 Photo by: Jin Dongmei

叶背 Leaf back
摄影：陈彬 Photo by: Chen Bin

花序 Inflorescence
摄影：陈彬 Photo by: Chen Bin

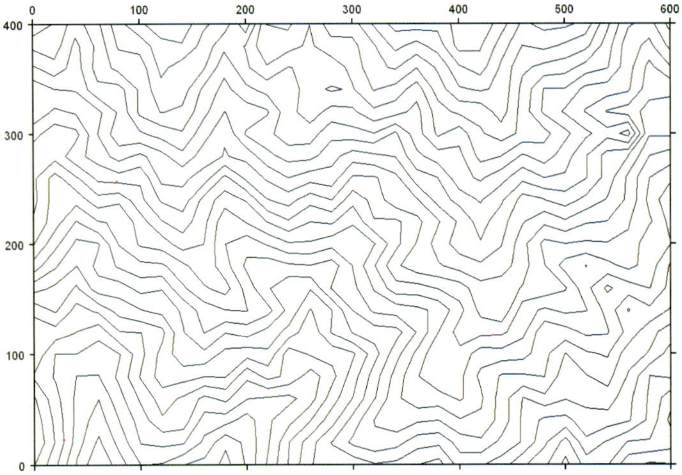

个体分布图 Distribution of Individuals

径级分布表 DBH Class

胸径区间 (Diameter class) (cm)	个体数 (No. of individuals in the plot)	比例 (Proportion)
1～2	2	1.00
2～5	0	0.00
5～10	0	0.00
10～20	0	0.00
20～30	0	0.00
30～60	0	0.00
≥60	0	0.00

59 臭辣吴萸（臭辣树） chòulàwúyú(chòulàshù) | Farges Evodia

Evodia fargesii Dode
芸香科 | Rutaceae

代码（SpCode）= EVOFAR
个体数（Individual number/24ha）= 12
最大胸径（Max DBH）= 22 cm
重要值排序（Importance value rank）= 113

落叶乔木，高达17m。树皮平滑，暗灰色。嫩枝紫褐色，散生小皮孔。叶5～9小叶，很少到11片。花序顶生。成熟心皮3～5个，紫红色。花期6～8月，果期8～10月。

Deciduous tree, up to 17 m tall. Bark smooth, dark grey. Young twigs purple brown, with scattered lenticels. Leaves 5-9-foliolate, rarely to 11. Inflorescence terminal. Follicles 3-5, purplish red when matured. Fl. Jun.-Aug, fr. Aug.-Oct..

花序 Inflorescence
摄影：方腾 Photo by: Fang Teng

幼苗 Seedling
摄影：方腾 Photo by: Fang Teng

果枝 Fruiting branch
摄影：方腾 Photo by: Fang Teng

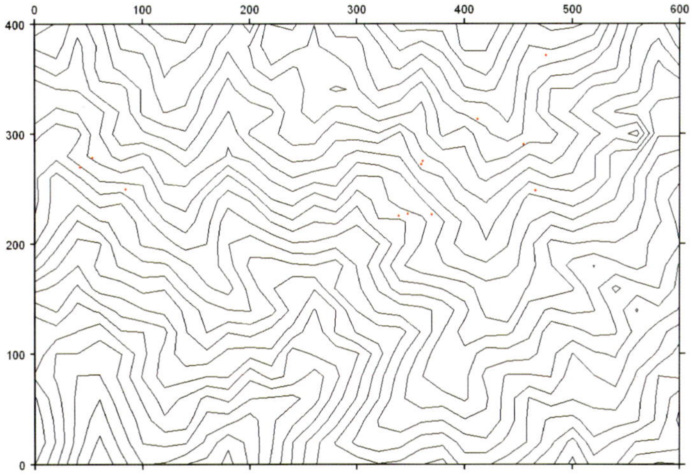

个体分布图 Distribution of Individuals

径级分布表 DBH Class

胸径区间 (Diameter class) (cm)	个体数 (No. of individuals in the plot)	比例 (Proportion)
1～2	0	0.00
2～5	4	0.33
5～10	2	0.17
10～20	3	0.25
20～30	3	0.25
30～60	0	0.00
≥60	0	0.00

60 岭南花椒

lǐngnánhuājiāo | Guangdong Prickleyash

Zanthoxylum austrosinense C. C. Huang
芸香科 | Rutaceae

代码（SpCode）= ZANAUS
个体数（Individual number/24ha）= 3
最大胸径（Max DBH）= 3 cm
重要值排序（Importance value rank）= 137

落叶灌木或小乔木，高达3m。雌雄异株，杂性同株。枝褐黑色。叶片具小叶5～11片。花序顶生，通常生于侧枝顶上。果梗暗紫红色，1～2cm。分果瓣黑紫色，种子直径约4mm，长3～4mm。花期3～4月，果期8～9月。

Deciduous shrub or small tree, up to 3 m tall, dioecious or rarely polygamo-monoecious, all parts glabrous. Branchlets blackish brown. Leaves 5-11-foliolate. Inflorescences terminal, usually on lateral branchlets. Fruit pedicel dark purplish red, 1-2 cm; follicles dark purplish red. Seeds ca. 4 mm × 3-4 mm. Fl. Mar.-Apr., fr. Aug.-Sep..

幼苗 Seedling
摄影：杨波 Photo by: Yang Bo

叶背 Leaf back
摄影：杨波 Photo by: Yang Bo

叶面的刺 Prickling on leaf
摄影：杨波 Photo by: Yang Bo

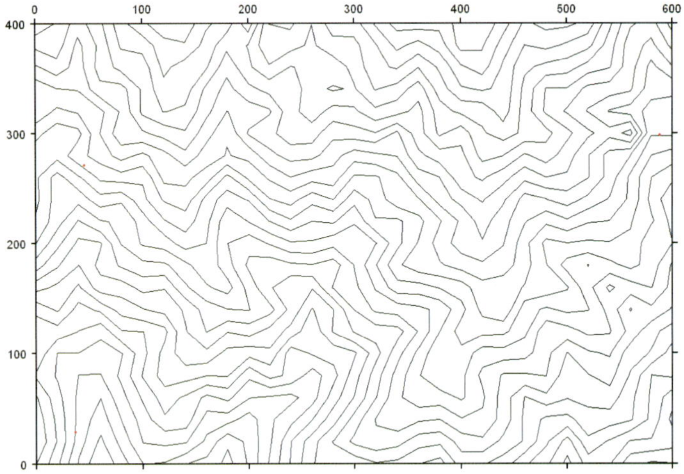

个体分布图 Distribution of Individuals

径级分布表 DBH Class

胸径区间 (Diameter class) (cm)	个体数 (No. of individuals in the plot)	比例 (Proportion)
1～2	2	0.67
2～5	1	0.33
5～10	0	0.00
10～20	0	0.00
20～30	0	0.00
30～60	0	0.00
≥60	0	0.00

61 苦树

kǔshù | India Quassia

Picrasma quassioides (D. Don) Benn.

苦木科 | Simaroubaceae

代码（SpCode）= PICQUA

个体数（Individual number/24ha）= 6

最大胸径（Max DBH）= 24.5 cm

重要值排序（Importance value rank）= 131

落叶乔木，高达10m。树皮紫褐色，平滑，有灰色斑纹，全株有苦味。叶互生，有小叶9～15。花雌雄异株，组成腋生复聚伞花序。心皮4～5，分离。核果成熟后蓝绿色。花期4～5月，果期6～9月。

Deciduous tree, up to 10 m tall; bark purplish brown, smooth, with gray stripes. Leaves alternate, 9-15-foliolate. Flowers dioecious, in axillary cymes. Carpels 4 or 5, free. Drupe blue-green when ripe. Fl. Apr.-May, fr. Jun.-Sep..

幼苗 Seedling
摄影：方腾 Photo by: Fang Teng

花序 Inflorescence
摄影：方腾 Photo by: Fang Teng

果序 Infructescence
摄影：陈彬 Photo by: Chen Bin

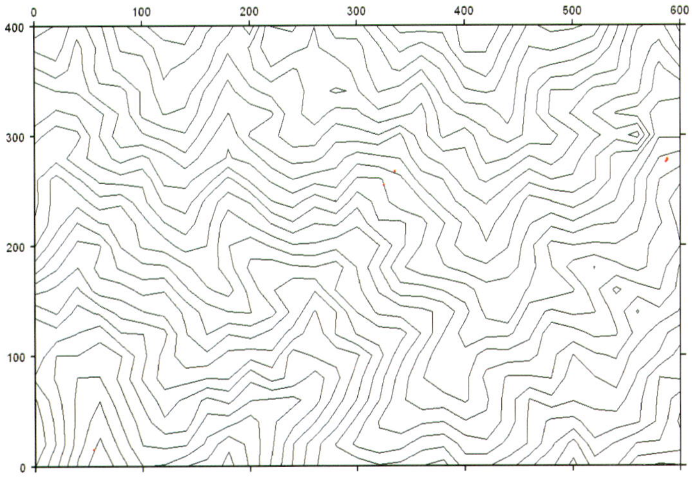

个体分布图 Distribution of Individuals

径级分布表 DBH Class

胸径区间 (Diameter class) (cm)	个体数 (No. of individuals in the plot)	比例 (Proportion)
1～2	1	0.17
2～5	1	0.17
5～10	2	0.33
10～20	1	0.17
20～30	1	0.17
30～60	0	0.00
≥60	0	0.00

62 酸味子（日本五月茶） suānwèizǐ(rìběnwǔyuèchá) | Japan Maytea

Antidesma japonicum Sieb. et Zucc.
大戟科 | Euphorbiaceae

代码（SpCode）= ANTJAP
个体数（Individual number/24ha）= 13
最大胸径（Max DBH）= 2.3 cm
重要值排序（Importance value rank）= 120

常绿灌木或小乔木，高2～8m。小枝纤细，幼时有柔毛，后变无毛。叶片纸质至近革质，侧脉5～12对。花序顶生或腋生，纤细，花单性同株。核果椭圆形。花期4～6月，果期7～9月。

Everngreen shrub or small tree 2-8 m tall; young twigs delicate, pubescent when young, glabrescent. Leaf blade papery to subleathery, lateral veins 5-12 pairs. Inflorescences terminal or axillary, slender, separate flower. Drupes ellipsoid. Fl. Apr.-Jun., fr. Jul.-Sep..

小枝 Branchlet
摄影：宋凯 Photo by: Song Kai

幼苗 Seedling
摄影：方腾 Photo by: Fang Teng

果枝 Fruiting branch
摄影：陈彤 Photo by: Chen Tong

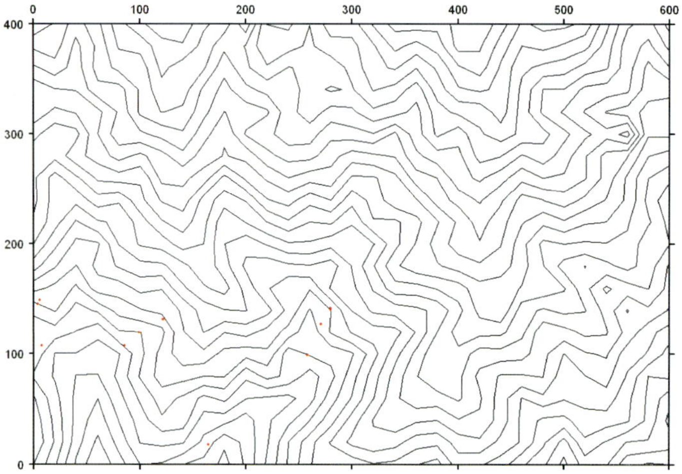

个体分布图 Distribution of Individuals

径级分布表 DBH Class

胸径区间 (Diameter class) (cm)	个体数 (No. of individuals in the plot)	比例 (Proportion)
1～2	12	0.92
2～5	1	0.08
5～10	0	0.00
10～20	0	0.00
20～30	0	0.00
30～60	0	0.00
≥60	0	0.00

63 算盘子 suànpánzi | Puberulous Glochidion

Glochidion puberum (L.) Hutch.
大戟科 | Euphorbiaceae

代码（SpCode）= GLOPUB
个体数（Individual number/24ha）= 14
最大胸径（Max DBH）= 4.3 cm
重要值排序（Importance value rank）= 109

落叶灌木至乔木，高达15m。雌雄同株，多分枝。小枝灰褐色密被柔毛。叶片纸质或近革质。花2~5朵簇生于叶腋。蒴果扁球形，密被毛，熟时红色。花期4~8月，果期7~11月。

Deciduous shrub to tree, up to 15 m tall; monoecious, much branched; branchlets gray-brown, densely pubescent. Leaf blade papery or subleathery. Flowers in axillary clusters, 2-5-flowered. Capsules depressed-globose, densely pubescent, reddish when mature. Fl. Apr.-Aug., fr. Jul.-Nov..

蒴果 Capsule
摄影：方腾 Photo by: Fang Teng

幼苗 Seedling
摄影：金冬梅 Photo by: Jin Dongmei

种子 Seeds
摄影：方腾 Photo by: Fang Teng

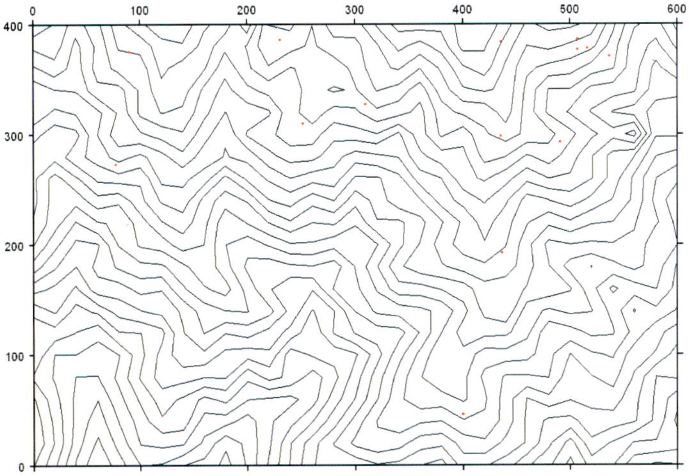

个体分布图 Distribution of Individuals

径级分布表 DBH Class

胸径区间 (Diameter class) (cm)	个体数 (No. of individuals in the plot)	比例 (Proportion)
1~2	5	0.36
2~5	9	0.64
5~10	0	0.00
10~20	0	0.00
20~30	0	0.00
30~60	0	0.00
≥60	0	0.00

64 青灰叶下珠

qīnghuīyèxiàzhū | Greyblue Underleaf pearl

Phyllanthus glaucus Wall. ex Müll. Arg.

大戟科 | Euphorbiaceae

代码（SpCode）= PHYGLA

个体数（Individual number/24ha）= 2

最大胸径（Max DBH）= 2 cm

重要值排序（Importance value rank）= 143

落叶灌木，高达8m，雌雄同株。全株光滑无毛。小枝纤细。叶片纸质。花丛生叶腋，经常有多朵雄花与一朵雌花。浆果球形到卵形，直径约1cm，黑紫色。花期4～6月，果期7～8月。

Deciduous shrub, up to 8 m tall, monoecious, glabrous throughout; branchlets delicate. Leaf blade papery. Inflorescence an axillary fascicle, usually with several male and 1 female flower. Fruit a berry, globose to oblate, ca. 1 cm in diam, black-purple. Fl. Apr.-Jun., fr. Jul.-Aug..

枝叶 Branch and leaves

摄影：方腾 Photo by: Fang Teng

花序 Inflorescence

摄影：赖阳均 Photo by: Lai Yangjun

果枝 Fruiting branch

摄影：陈彬 Photo by: Chen Bin

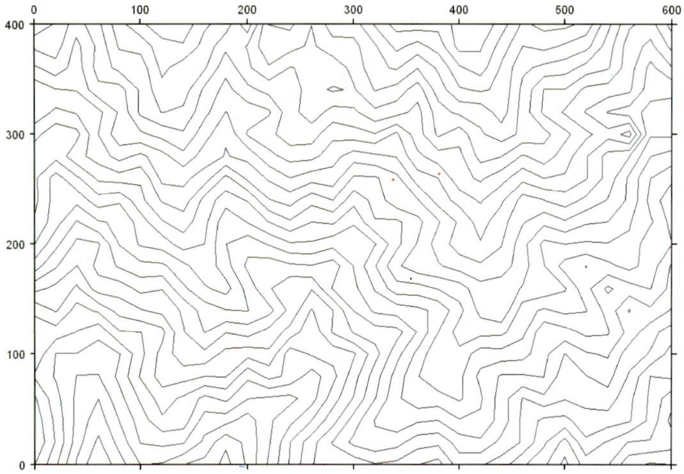

个体分布图 Distribution of Individuals

径级分布表 DBH Class

胸径区间 (Diameter class) (cm)	个体数 (No. of individuals in the plot)	比例 (Proportion)
1～2	2	1.00
2～5	0	0.00
5～10	0	0.00
10～20	0	0.00
20～30	0	0.00
30～60	0	0.00
≥60	0	0.00

65 白木乌桕（白乳木） báimùwūjiù(báirǔmù) | Japan Tallowtree

Sapium japonica (Sieb. et Zucc.) Pax et Hoffm.
大戟科 | Euphorbiaceae

代码（SpCode）= SAPJAP
个体数（Individual number/24ha）= 5
最大胸径（Max DBH）= 6.8 cm
重要值排序（Importance value rank）= 132

落叶灌木或乔木，高达8m。全株光滑无毛。小枝纤细，光滑，灰褐色。叶片纸质，在基部有2腺体。花序顶生。蒴果三角状球形。种子有茶褐色斑纹。花期5～6月，果期7～9月。

Deciduous shrub or small tree, up to 8 m tall, glabrous; branches slender, smooth, gray-brown. Leaf blade papery, with 2 glands at base of midrib. Inflorescences terminal. Capsules triangular-globose. Seeds with tawny maculate stripes. Fl. May-Jun., fr. Jul.-Sep..

小枝 Branchlet
摄影：方腾 Photo by: Fang Teng

花序 Inflorescence
摄影：方腾 Photo by: Fang Teng

蒴果 Capsule
摄影：方腾 Photo by: Fang Teng

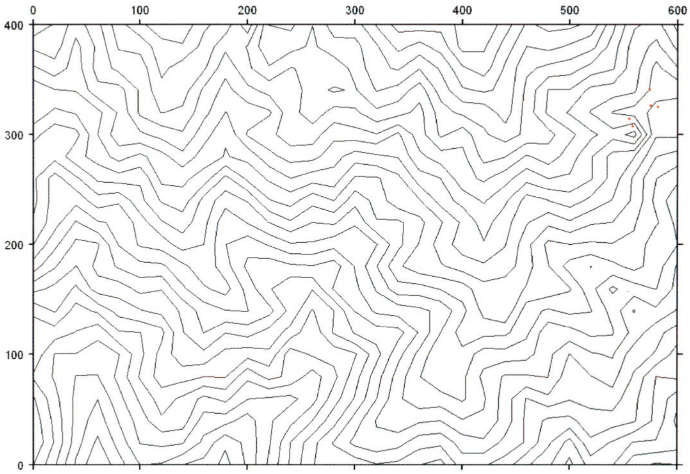

个体分布图 Distribution of Individuals

径级分布表 DBH Class

胸径区间 (Diameter class) (cm)	个体数 (No. of individuals in the plot)	比例 (Proportion)
1～2	1	0.20
2～5	1	0.20
5～10	3	0.60
10～20	0	0.00
20～30	0	0.00
30～60	0	0.00
≥60	0	0.00

66 木油桐

mùyóutóng | Muyou Oiltung

Vernicia montana Lour.

大戟科 | Euphorbiaceae

代码（SpCode）= VERMON

个体数（Individual number/24ha）= 1

最大胸径（Max DBH）= 1.1 cm

重要值排序（Importance value rank）= 158

落叶乔木，高达 20m。常雌雄异株。树皮褐色，小枝无毛，具稀疏的突起皮孔。叶阔卵形，长 8 ~ 20cm，宽 6 ~ 18cm。花与叶同放，常单性。核果卵球形，有种子 3 颗。花期 4 ~ 5 月，果期 7 ~ 8 月。

Deciduous tree, up to 20 m tall, usually dioecious; bark brown; branches glabrous, with sparsely elevated lenticels. Leaf blade broadly ovate, 8-20 cm × 6-18 cm. Inflorescences produced with new leaves, usually unisexual. Drupes ovoid, 3-seeded. Fl. Apr. -May., fr. Jul.-Oct..

叶 Leaf

摄影：宋凯 Photo by: Song Kai

雄花序 Male inflorescence

摄影：方腾 Photo by: Fang Teng

蒴果 Capsule

摄影：陈彬 Photo by: Chen Bin

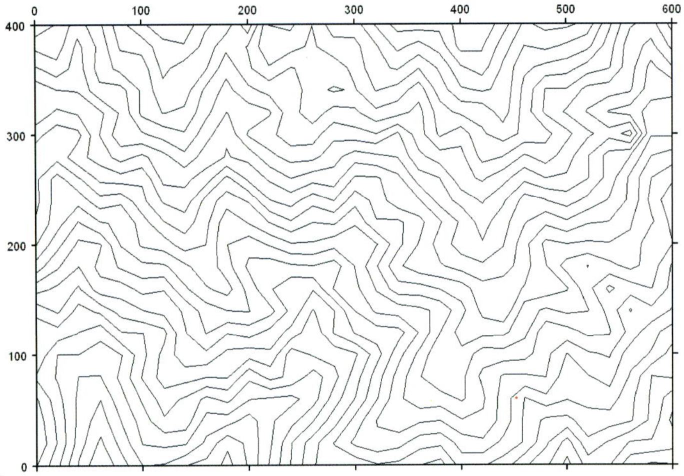

个体分布图 Distribution of Individuals

径级分布表 DBH Class

胸径区间 (Diameter class) (cm)	个体数 (No. of individuals in the plot)	比例 (Proportion)
1～2	1	1.00
2～5	0	0.00
5～10	0	0.00
10～20	0	0.00
20～30	0	0.00
30～60	0	0.00
≥60	0	0.00

67 虎皮楠 hǔpínán | Oldham Tigernanmu

Daphniphyllum oldhamii (Hemsl.) Rosenthal
虎皮楠科 | Daphniphyllaceae

代码（SpCode）= DAPOLD
个体数（Individual number/24ha）= 2716
最大胸径（Max DBH）= 40.7 cm
重要值排序（Importance value rank）= 8

常绿灌木或乔木，高4～15m。小枝纤细，暗褐色。叶纸质，叶背粉白色。侧脉8～15对，网脉两面显著突起。花单性同株。核果椭圆形或倒卵圆形，暗褐色至黑色。花期3～5月，果期8～11月。

Evergreen shrub or tree, 4-15 m tall; branchlets slender, dark brown. Leaf blade chartaceous, abaxially whitish; lateral veins 8-15 pairs, slender, prominent on both surfaces. Flowers unisexual. Drupe ellipsoidal or obovate-globose, dark brown to black. Fl. Mar.-May, fr. Aug.-Nov..

幼苗 Seedling
摄影：金冬梅 Photo by: Jin Dongmei

叶 Leaf
摄影：喻勋林 Photo by: Yu Xunlin

雌花序 Female inflorescence
摄影：方腾 Photo by: Fang Teng

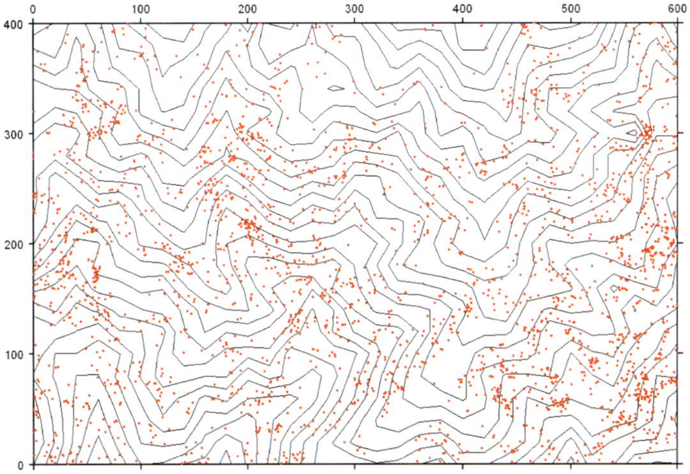

个体分布图 Distribution of Individuals

径级分布表 DBH Class

胸径区间 (Diameter class) (cm)	个体数 (No. of individuals in the plot)	比例 (Proportion)
1～2	725	0.27
2～5	776	0.29
5～10	520	0.19
10～20	563	0.21
20～30	123	0.05
30～60	11	0.00
≥60	0	0.00

68 南酸枣

nánsuānzǎo | Southern Wildjujube

Choerospondias axillaris (Roxb.) B. L. Burtt et A. W. Hill
漆树科 | Anacardiaceae

代码（SpCode）= CHOAXI
个体数（Individual number/24ha）= 16
最大胸径（Max DBH）= 33 cm
重要值排序（Importance value rank）= 107

落叶乔木，高8~20m。小枝暗紫褐色，被微柔毛至无毛，具皮孔。奇数羽状复叶。核果成熟时黄色，果核长2~2.5cm，径1.2~1.5cm，顶端具5个小孔。花期4~5月，果期9~11月。

Deciduous tree, 8-20 m tall; branchlets dark purplish brown, minutely pubescent to glabrous, lenticellate. Leaf blade imparipinnately compound. Drupe yellow at maturity; endocarp 2-2.5 cm × 1.2-1.5 cm, with 5 eyelets at apex. Fl. Apr.-May, fr. Sep.-Nov..

植株 Whole plant
摄影：陈彬 Photo by: Chen Bin

小枝 Branchlet
摄影：陈国科 Photo by: Chen Guoke

核果 Drupe
摄影：陈国科 Photo by: Chen Guoke

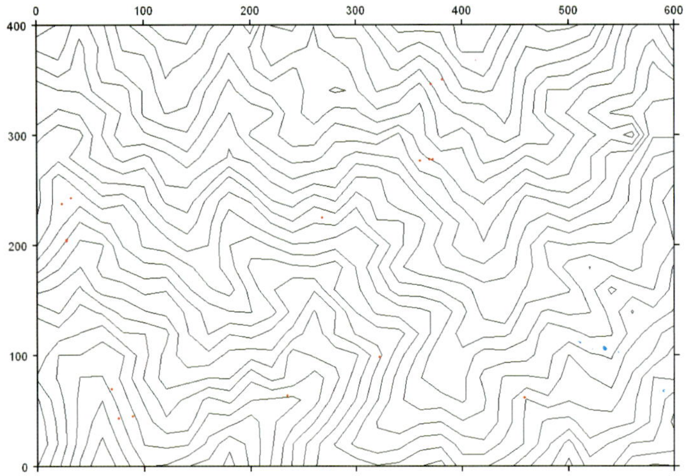

个体分布图 Distribution of Individuals

径级分布表 DBH Class

胸径区间 (Diameter class) (cm)	个体数 (No. of individuals in the plot)	比例 (Proportion)
1~2	2	0.13
2~5	1	0.06
5~10	4	0.25
10~20	3	0.19
20~30	3	0.19
30~60	3	0.19
≥60	0	0.00

69 白背麸杨 báibèifūyáng | Whiteback Sumac

Rhus hypoleuca Champ. Ex Benth.
漆树科 | Anacardiaceae

代码（SpCode）= RHUHYP
个体数（Individual number/24ha）= 199
最大胸径（Max DBH）= 12.7 cm
重要值排序（Importance value rank）= 66

落叶灌木或小乔木。小枝紫褐色，幼时被灰色微绒毛后变无毛，疏生皮孔。奇数羽状复叶，小叶无柄，9～17，纸质。圆锥花序顶生。核果被白色长柔毛和红色腺毛。花期8月，果期9～10月。

Deciduous shrub or small tree; branchlets minutely grayish pubescent to glabrous, sparsely lenticellate. Leaf blade imparipinnately compound, leaflets sessile, 9-17, papery. Inflorescence paniculate, terminal. Drupe mixed pilose and red glandular-pubescent. Fl. Aug., fr. Sep.-Oct..

果序 Infructescence
摄影：方腾 Photo by: Fang Teng

叶背 Leaf back
摄影：陈彬 Photo by: Chen Bin

花序 Inflorescence
摄影：方腾 Photo by: Fang Teng

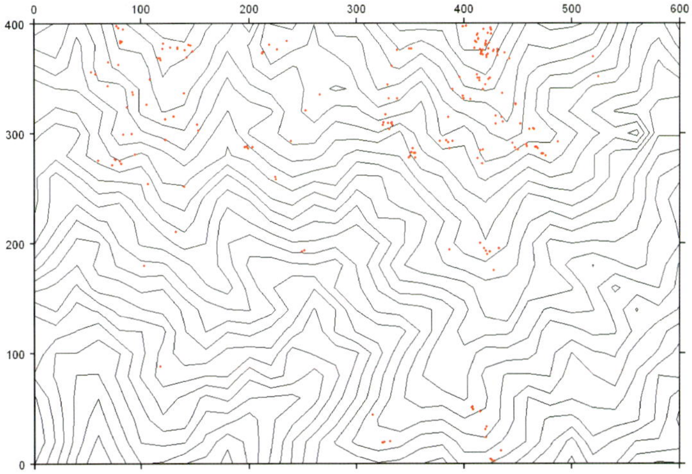

个体分布图 Distribution of Individuals

径级分布表 DBH Class

胸径区间 (Diameter class) (cm)	个体数 (No. of individuals in the plot)	比例 (Proportion)
1～2	27	0.14
2～5	89	0.45
5～10	80	0.40
10～20	3	0.02
20～30	0	0.00
30～60	0	0.00
≥60	0	0.00

70 野漆

yěqī | Field Lacquertree

Toxicodendron succedaneum (L.) Kuntze

漆树科 | Anacardiaceae

代码（SpCode）= TOXSUC

个体数（Individual number/24ha）= 335

最大胸径（Max DBH）= 36.2 cm

重要值排序（Importance value rank）= 38

落叶乔木或灌木，高 2 ~ 10m。小枝粗壮。奇数羽状复叶，常集生小枝顶端。小叶 5 ~ 15，叶背常具白粉。圆锥花序顶生，多分枝。核果大，偏斜，压扁，淡黄色，无毛。花期 5 ~ 6 月，果期 10 月。

Deciduous tree or shrub, 2-10 m tall; branchlets stout. Leaf blade imparipinnately compound, leaflets 5-15, glaucous abaxially. Inflorescence paniculate, terminal, many branched. Drupe large, asymmetrical, compressed, yellow, glabrous; mesocarp thick, white, waxy, with brown longitudinal resin ducts. Fl. May-Jun., fr. Oct..

幼苗 Seedling

摄影：金冬梅 Photo by: Jin Dongmei

小枝 Branchlet

摄影：方腾 Photo by: Fang Teng

核果 Drupe

摄影：陈国科 Photo by: Chen Guoke

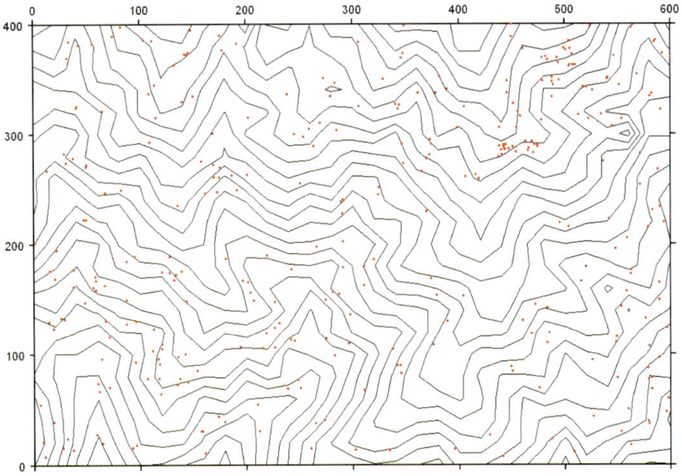

个体分布图 Distribution of Individuals

径级分布表 DBH Class

胸径区间 (Diameter class) (cm)	个体数 (No. of individuals in the plot)	比例 (Proportion)
1～2	79	0.24
2～5	81	0.24
5～10	54	0.16
10～20	57	0.17
20～30	51	0.15
30～60	13	0.04
⩾60	0	0.00

71 厚叶冬青 hòuyèdōngqīng | Thickleaf Holly

Ilex elmerrilliana S. Y. Hu
冬青科 | Aquifoliaceae

代码（SpCode）= ILEELM
个体数（Individual number/24ha）= 402
最大胸径（Max DBH）= 18.3 cm
重要值排序（Importance value rank）= 39

常绿灌木或小乔木。高2～7m。树皮灰褐色。当年生枝红褐色，具纵棱脊，无毛。叶片厚革质，两面无毛。花簇生于二年生与当年生枝叶腋。核果球形，红色，具分核6～7，花期4～5月，果期7～11月。

Evergreen shrub or small tree, 2-7 m tall; bark gray-brown. First year's branchlets red-brown, longitudinally angular and ridged, glabrous; leaf blade thickly leathery, both surfaces glabrous. Inflorescences fasciculate, axillary on second or first year's branchlets. Fruit red, globose, pyrenes 6 or 7. Fl. Apr.-May, fr. Jul.-Nov..

幼苗 Seedling
摄影：金冬梅 Photo by: Jin Dongmei

枝叶 Branch and leaves
摄影：陈建华 Photo by: Chen Jianhua

花枝 Flowering branch
摄影：方腾 Photo by: Fang Teng

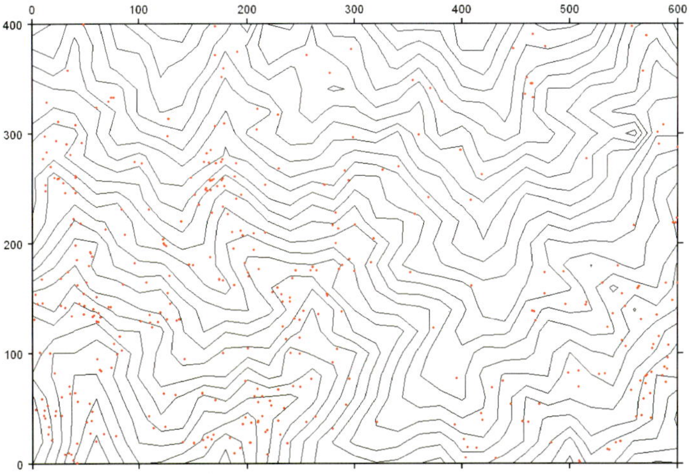

个体分布图 Distribution of Individuals

径级分布表 DBH Class

胸径区间 (Diameter class) (cm)	个体数 (No. of individuals in the plot)	比例 (Proportion)
1～2	178	0.44
2～5	136	0.34
5～10	66	0.16
10～20	22	0.05
20～30	0	0.00
30～60	0	0.00
≥60	0	0.00

72 短梗冬青

duǎngěngdōngqīng | Short-pedicel Holly

Ilex buergeri Miq.

冬青科 | Aquifoliaceae

代码（SpCode）= ILEBUE
个体数（Individual number/24ha）= 409
最大胸径（Max DBH）= 24.2 cm
重要值排序（Importance value rank）= 43

常绿灌木或乔木，高1～15m。小枝圆柱形，具纵棱脊和槽，密被短柔毛，老枝变无毛。叶片革质，侧脉7～8对，具疏而不规则的浅锯齿。聚伞花序簇生于叶腋内。果球形或近球形，熟时红色，表面具小瘤点。分核4。花期4～6月，果期10～11月。

Evergreen shrub or small tree, 1-15 m tall. Branchlets terete, densely pubescent; old branchlets glabrescent. Leaf blade leathery, lateral veins 7-8 pairs. Inflorescences: 1-flowered cymes, fasciculate, fascicles 4-10-flowered, axillary on branchlets. Fruit red, globose or subglobose, tuberculate; pyrenes 4. Fl. Apr.-Jun., fr. Oct.-Nov..

幼果 Fruit(immature)
摄影：方腾 Photo by: Fang Teng

小枝 Branchlet
摄影：陈国科 Photo by: Chen Guoke

花枝 Flowering branch
摄影：刘晓娟 Photo by: Liu Xiaojuan

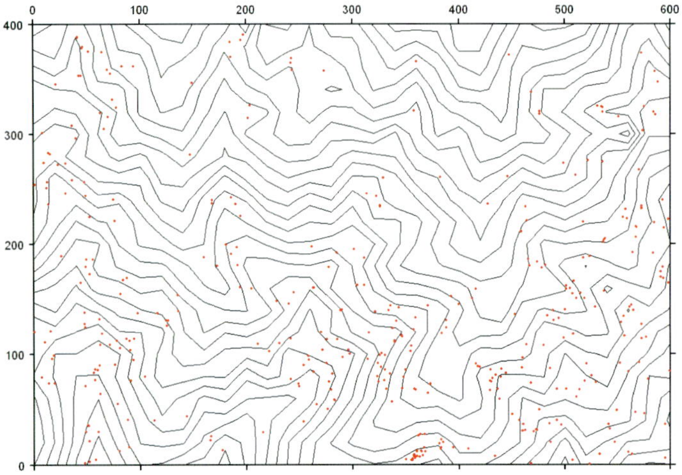

个体分布图 Distribution of Individuals

径级分布表 DBH Class

胸径区间 (Diameter class) (cm)	个体数 (No. of individuals in the plot)	比例 (Proportion)
1～2	270	0.66
2～5	113	0.28
5～10	14	0.03
10～20	10	0.02
20～30	2	0.00
30～60	0	0.00
≥60	0	0.00

73 大叶冬青 dàyèdōngqīng | Broadleaf Holly

Ilex latifolia Thunb.
冬青科 | Aquifoliaceae

代码（SpCode）= ILELAT
个体数（Individual number/24ha）= 80
最大胸径（Max DBH）= 29.8 cm
重要值排序（Importance value rank）= 71

常绿乔木，高达 20m。全株无毛。树皮灰黑色，光滑，分枝粗壮，具纵棱及槽。叶片厚革质，长达 15cm。花序生于二年生枝的叶腋，无总梗。果红色，球形，分核 4。花期 4 月，果期 9 ~ 10 月。

Evergreen tree, up to 20 m tall, glabrous throughout; bark gray-black, smooth. Branches strong, longitudinally ridged and sulcate; leaf blade thickly leathery, always longer than 15cm. Inflorescences axillary on second year's branchlets. Fruit red, globose; pyrenes 4. Fl. Apr., fr. Sep.-Oct..

幼苗 Seedling
摄影：方腾 Photo by: Fang Teng

花 Flower
摄影：陈彬 Photo by: Chen Bin

果枝 Fruiting branch
摄影：陈彬 Photo by: Chen Bin

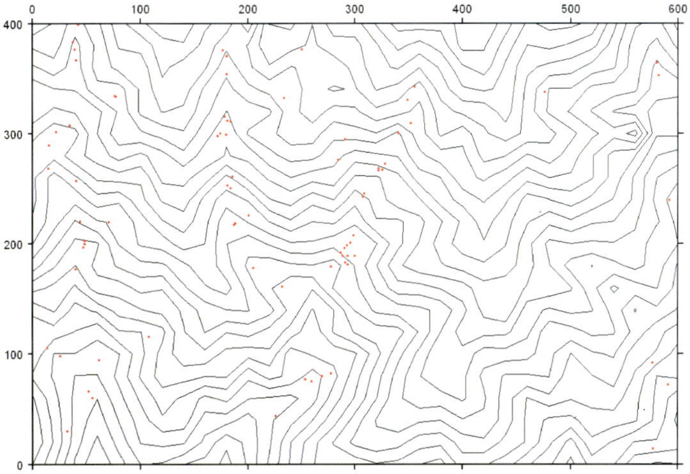

个体分布图 Distribution of Individuals

径级分布表 DBH Class

胸径区间 (Diameter class) (cm)	个体数 (No. of individuals in the plot)	比例 (Proportion)
1～2	62	0.78
2～5	11	0.14
5～10	4	0.05
10～20	2	0.03
20～30	1	0.01
30～60	0	0.00
≥60	0	0.00

74 木姜叶冬青

mùjiāngyèdōngqīng | Litsealeaf Holly

Ilex litseaefolia Hu et Tang
冬青科 | Aquifoliaceae

代码（SpCode）= ILELIT
个体数（Individual number/24ha）= 156
最大胸径（Max DBH）= 16 cm
重要值排序（Importance value rank）= 85

常绿灌木至小乔木，高达12m。当年生幼枝具细棱。顶芽圆锥形，被棕黄色长柔毛。叶片革质。花生于当年生枝叶腋。果红色，球形，分核5。花期5～6月，果期8～11月。

Evergreen shrub to small tree, up to 12 m tall. First year branchlets fine longitudinally ridged; terminal bud cone-shaped, with yellowish brown villus. Leaf blade leathery. Inflorescence axillary in first year branchlets. Fruit red, globose; pyrenes 5. Fl. May-Jun., fr. Aug.-Nov..

植株 Whole plant
摄影：方腾 Photo by: Fang Teng

花 Flower
摄影：方腾 Photo by: Fang Teng

浆果状核果 Berrylike drupe
摄影：方腾 Photo by: Fang Teng

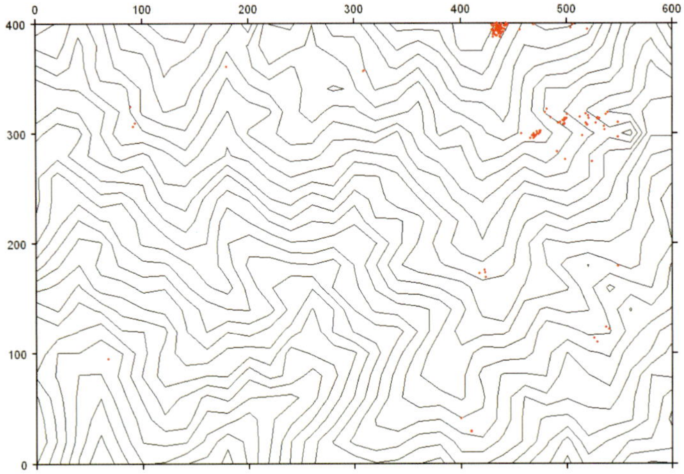

个体分布图 Distribution of Individuals

径级分布表 DBH Class

胸径区间 (Diameter class) (cm)	个体数 (No. of individuals in the plot)	比例 (Proportion)
1～2	41	0.26
2～5	60	0.38
5～10	37	0.24
10～20	18	0.12
20～30	0	0.00
30～60	0	0.00
≥60	0	0.00

75 小果冬青 xiǎoguǒdōngqīng | Smallfruit Holly

Ilex micrococca Maxim.
冬青科 | Aquifoliaceae

代码（SpCode）= ILEMIC
个体数（Individual number/24ha）= 26
最大胸径（Max DBH）= 55.8 cm
重要值排序（Importance value rank）= 92

落叶乔木，高达20m。小枝粗壮，无毛，具显著的白色气孔。叶片膜质或纸质，侧脉5～8对。聚伞花序生于当年生枝的叶腋内。果实红色或黄色，直径约3mm，分核6～8。花期5～6月，果期9～10月。

Deciduous tree, up to 20 m tall. Branchlets thick, glabrous, with conspicuous, white lenticels. Leaf blade membranous or papery, lateral veins 5-8 pairs. Inflorescences: compound cymes, axillary on first year's branchlets. Fruit red or yellow, globose, ca. 3 mm in diam; pyrenes 6-8. Fl. May-Jun., fr. Sep.-Oct..

幼苗 Seedling
摄影：方腾 Photo by: Fang Teng

花枝 Flowering branch
摄影：方腾 Photo by: Fang Teng

果枝 Fruiting branch
摄影：方腾 Photo by: Fang Teng

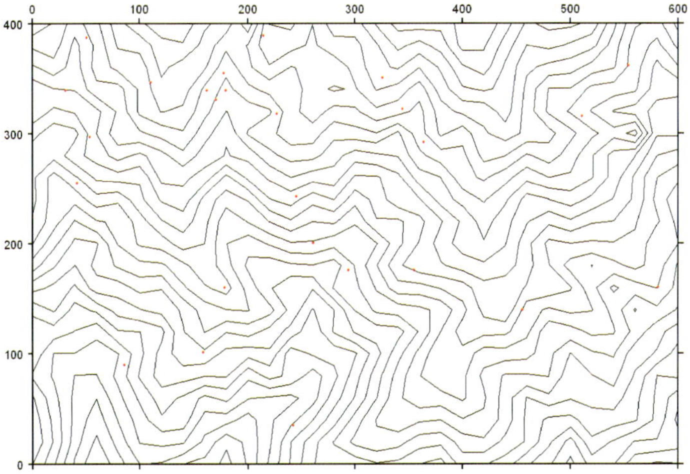

个体分布图 Distribution of Individuals

径级分布表 DBH Class

胸径区间 (Diameter class) (cm)	个体数 (No. of individuals in the plot)	比例 (Proportion)
1～2	3	0.12
2～5	4	0.15
5～10	2	0.08
10～20	3	0.12
20～30	4	0.15
30～60	10	0.38
≥60	0	0.00

76 毛冬青

máodōngqīng | Pilous Holly

Ilex pubescens Hook. & Arn.

冬青科 | Aquifoliaceae

代码（SpCode）= ILEPUB

个体数（Individual number/24ha）= 105

最大胸径（Max DBH）= 4.2 cm

重要值排序（Importance value rank）= 64

常绿灌木，高3～4m。小枝纤细，近四棱形，灰褐色，密被长硬毛，具纵棱脊，无皮孔。叶片纸质或膜质。花序簇生于1～2年生叶腋内。果红色，球形，分核5～7枚。花期4～5月，果期8～11月。

Evergreen shrub, 3-4 m tall. Branchlets slender, subquadrangular, gray-brown, densely hirsute or villous, longitudinally ridged and rugose; lenticels absent; leaf blade papery, or membranous. Inflorescences: cymes, axillary on first to second year's branchlets. Fruit red, globose, pyrenes 5-7. Fl. Apr.-May, fr. Aug.-Nov..

幼苗 Seedling

摄影：金冬梅 Photo by: Jin Dongmei

花 Flower

摄影：贾琪 Photo by: Jia Qi

果枝 Fruiting branch

摄影：张金龙 Photo by: Zhang Jinlong

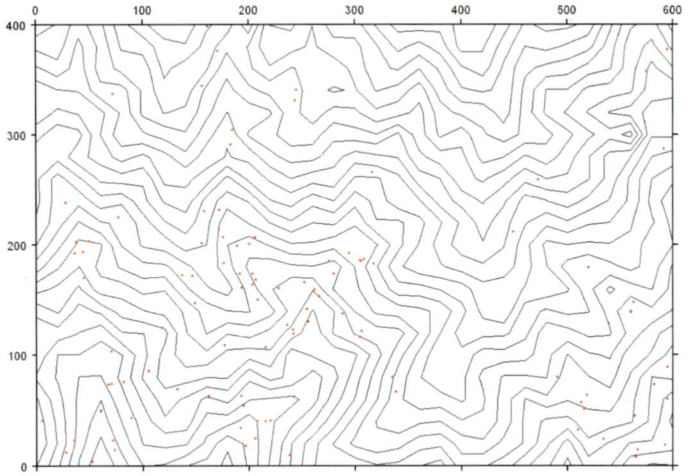

个体分布图 Distribution of Individuals

径级分布表 DBH Class

胸径区间 (Diameter class) (cm)	个体数 (No. of individuals in the plot)	比例 (Proportion)
1～2	89	0.85
2～5	16	0.15
5～10	0	0.00
10～20	0	0.00
20～30	0	0.00
30～60	0	0.00
≥60	0	0.00

77 冬青

dōngqīng | Chinese Holly

Ilex chinensis Sims
冬青科 | Aquifoliaceae

代码（SpCode）= ILECHI
个体数（Individual number/24ha）= 303
最大胸径（Max DBH）= 29.7 cm
重要值排序（Importance value rank）= 49

常绿乔木，高达13m。树皮灰黑色，当年生小枝浅灰色，圆柱形；老枝具不明显的小皮孔。叶片薄革质至革质。聚伞花序生于当年生枝叶腋。果红色，长球形，分核4～5。花期4～6月，果期7～12月。

Evergreen tree, up to 13 m tall; bark gray-black. First year's branchlets gray, terete; older branchlets' lenticels obscure, small. Leaf blade thinly leathery to leathery. Inflorescences: cymes, axillary on first year' s branchlets. Fruit red, narrowly globose, pyrenes 4 or 5. Fl. Apr.-Jun., fr. Jul.-Dec..

幼苗 Seedling
摄影：金冬梅 Photo by: Jin Dongmei

花序 Inflorescence
摄影：方腾 Photo by: Fang Teng

果枝 Fruiting branch
摄影：陈国科 Photo by: Chen Guoke

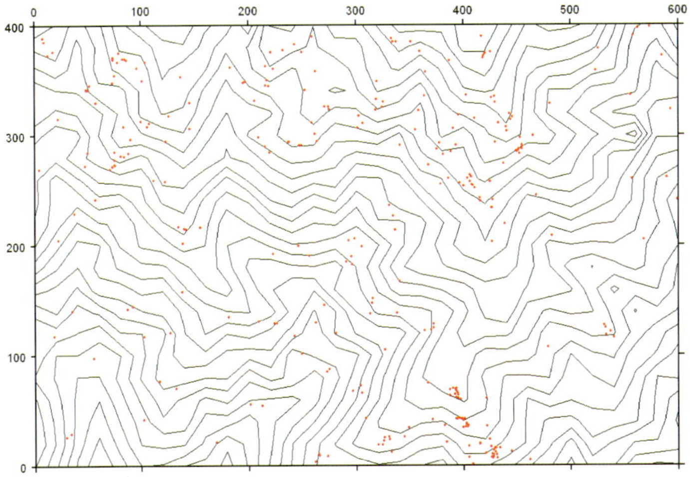

个体分布图 Distribution of Individuals

径级分布表 DBH Class

胸径区间 (Diameter class) (cm)	个体数 (No. of individuals in the plot)	比例 (Proportion)
1～2	123	0.41
2～5	119	0.39
5～10	43	0.14
10～20	15	0.05
20～30	3	0.01
30～60	0	0.00
≥60	0	0.00

78 铁冬青 tiědōngqīng | Iron Holly

Ilex rotunda Thunb.

冬青科 | Aquifoliaceae

代码（SpCode）= ILEROT

个体数（Individual number/24ha）= 197

最大胸径（Max DBH）= 21.3 cm

重要值排序（Importance value rank）= 52

常绿灌木或乔木，高达 20m。树皮灰色至灰黑色。叶仅见于当年生枝上。叶片薄革质或纸质。聚伞花序单生于当年生枝叶腋。果红色，近球形或稀椭圆形。分核 5 ~ 7 个。花期 4 月，果期 8 ~ 12 月。

Evergreen shrub or small tree, up to 20 m tall; bark gray to gray-black. Leaf only occurred on first year branchlets, blade thinly leathery or papery. Inflorescences: cymes, axillary on first year's branchlets. Fruit red, subglobose, rarely ellipsoid, pyrenes 5-7. Fl. Apr., fr. Aug.-Dec..

幼苗 Seedling

摄影：方腾 Photo by: Fang Teng

花序 Inflorescence

摄影：方腾 Photo by: Fang Teng

果枝 Fruiting branch

摄影：方腾 Photo by: Fang Teng

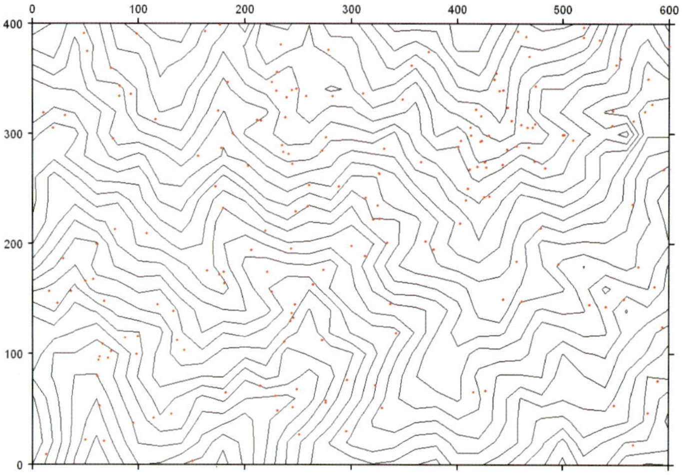

个体分布图 Distribution of Individuals

径级分布表 DBH Class

胸径区间 (Diameter class) (cm)	个体数 (No. of individuals in the plot)	比例 (Proportion)
1～2	92	0.47
2～5	65	0.33
5～10	25	0.13
10～20	12	0.06
20～30	3	0.02
30～60	0	0.00
≥60	0	0.00

79 香冬青 xiāngdōngqīng | Fragrant Holly

Ilex suaveolens (Lévl.) Loes.
冬青科 | Aquifoliaceae

代码（SpCode）= ILESUA
个体数（Individual number/24ha）= 55
最大胸径（Max DBH）= 30.6 cm
重要值排序（Importance value rank）= 77

常绿乔木，高达 15m。当年生小枝褐色，具棱角，秃净。二年生枝近圆柱形，皮孔椭圆形，隆起。叶片革质。伞形花序单生叶腋。果成熟时红色，近球形，分核 4。花期 4 ~ 6 月，果期 9 ~ 12 月。

Evergreen tree, up to 15 m tall. Young branchlets brown, angular, glabrous throughout; older branchlets subterete; lenticels raised, elliptic. Leaf blade leathery. Infructescences: cymes, solitary, axillary. Fruit red, narrowly globose; pyrenes 4. Fl. Apr.-Jun., fr. Sep.-Dec..

植株 Whole plant
摄影：刘晓娟 Photo by: Liu Xiaojuan

小枝 Branchlet
摄影：陈国科 Photo by: Chen Guoke

果枝 Fruiting branch
摄影：杜彦君 Photo by: Du Yanjun

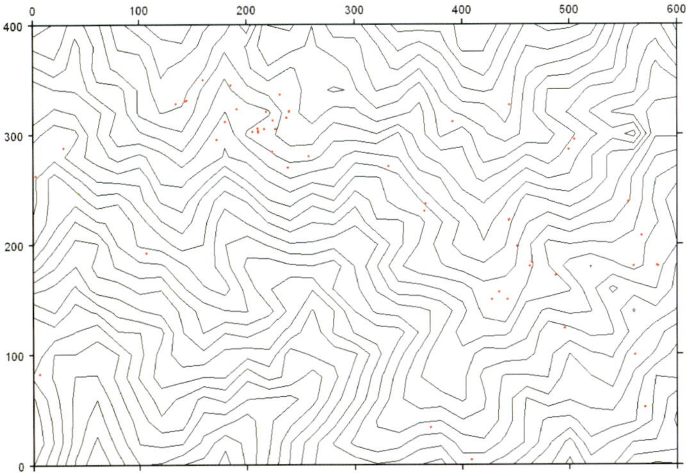

个体分布图 Distribution of Individuals

径级分布表 DBH Class

胸径区间 (Diameter class) (cm)	个体数 (No. of individuals in the plot)	比例 (Proportion)
1～2	21	0.38
2～5	19	0.35
5～10	12	0.22
10～20	0	0.00
20～30	2	0.04
30～60	1	0.02
⩾60	0	0.00

80 尾叶冬青

wěiyèdōngqīng | Wilson Holly

Ilex wilsonii Loes.
冬青科 | Aquifoliaceae

代码（SpCode）= ILEWIL
个体数（Individual number/24ha）= 31
最大胸径（Max DBH）= 10.9 cm
重要值排序（Importance value rank）= 91

常绿灌木或乔木，高2～10m。树皮灰白色，光滑。小枝平滑，无皮孔。当年生幼枝具纵棱沟，光滑。叶片厚革质。花序簇生于二年生枝的叶腋内。果红色，球形，分核4。花期5～6月，果期8～10月。

Evergreen shrub or small tree, 2-10 m tall; bark gray-white, smooth. Branchlets smooth, glabrous; lenticels absent; first year's branchlets longitudinally angular and sulcate, glabrous. Leaf blade thickly leathery. Inflorescences: cymes, fasciculate, axillary on second year's branchlets. Fruit red, globose; pyrenes 4. Fl. May-Jun., fr. Aug.-Oct..

果枝 Fruiting branchlet
摄影：方腾 Photo by: Fang Teng

花枝 Flowering branch
摄影：陈彬 Photo by: Chen Bin

果序 Infructescence
摄影：方腾 Photo by: Fang Teng

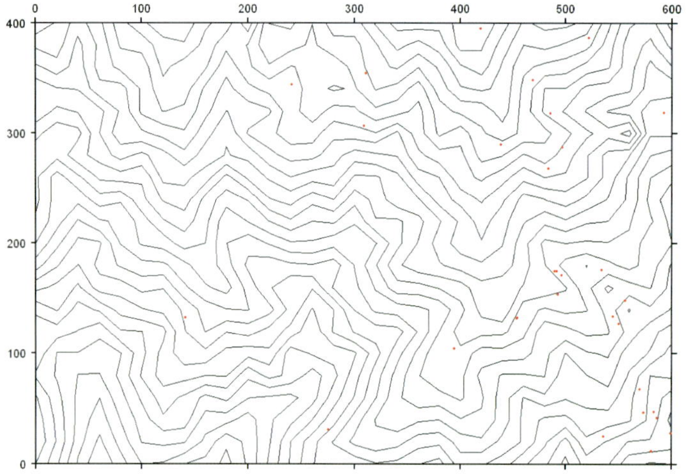

个体分布图 Distribution of Individuals

径级分布表 DBH Class

胸径区间 (Diameter class) (cm)	个体数 (No. of individuals in the plot)	比例 (Proportion)
1～2	8	0.26
2～5	11	0.35
5～10	11	0.35
10～20	1	0.03
20～30	0	0.00
30～60	0	0.00
≥60	0	0.00

81 中华卫矛（矩叶卫矛） zhōnghuáwèimáo(jǔyèwèimáo) | Chinese Euonymus

Euonymus nitidus Benth.
卫矛科 | Celastraceae

代码（SpCode）= EUONIT
个体数（Individual number/24ha）= 14
最大胸径（Max DBH）= 15.8 cm
重要值排序（Importance value rank）= 114

常绿灌木或小乔木，高 2 ~ 10m。叶革质，质地坚实。聚伞花序多花，1 ~ 4 次假二叉分枝，花序梗及分枝均较细长。蒴果四角卵圆形，略带紫色或红色；种子假种皮橙黄色。花期 3 ~ 7 月，果期 7 月至翌年 1 月。

Evergreen shrub to small tree, 2-10 m tall. Leaf blade leathery to thickly papery, taxture solid. Peduncle 1-4 × dichotomously branched, slender, with many flowers. Capsule tetra-globose, pinkish or reddish when fresh. Aril of seeds orange. Fl. Mar.-Jul., fr. Jul. to Jan. of next year.

叶 Leaf
摄影：张金龙 Photo by: Zhang Jinlong

叶背 Leaf back
摄影：陈国科 Photo by: Chen Guoke

果枝 Fruiting branch
摄影：张金龙 Photo by: Zhang Jinlong

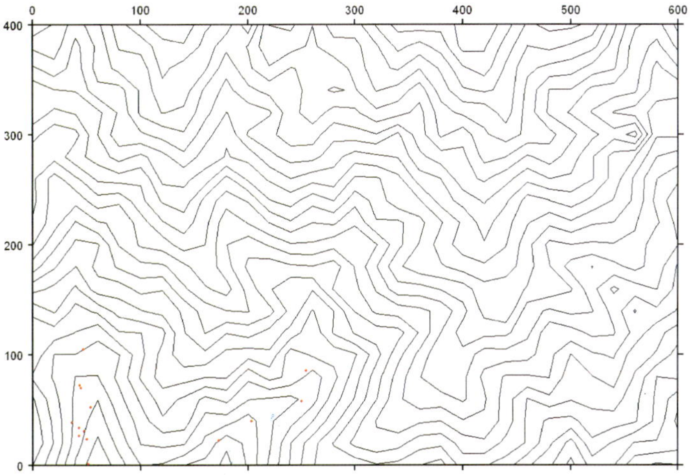

个体分布图 Distribution of Individuals

径级分布表 DBH Class

胸径区间 (Diameter class) (cm)	个体数 (No. of individuals in the plot)	比例 (Proportion)
1～2	3	0.21
2～5	4	0.29
5～10	2	0.14
10～20	5	0.36
20～30	0	0.00
30～60	0	0.00
≥60	0	0.00

82 大果卫矛 dàguǒwèimáo | Largefruit Euonymus

Euonymus myrianthus Hemsl.

卫矛科 | Celastraceae

代码（SpCode）= EUOMYR

个体数（Individual number/24ha）= 1

最大胸径（Max DBH）= 9.4 cm

重要值排序（Importance value rank）= 151

常绿灌木到小乔木，高3～12m。叶厚革质。侧脉7～9对。聚伞花序1或2次分枝。花黄色，直径达10mm。蒴果倒卵状，具四棱，带红色或紫色。种子2～4，假种皮橘黄色。花期4～6月，果期8～9月。

Evergreen shrub to small tree, 3-12 m tall. Leaf blade thickly leathery; lateral veins 7-9 pairs. Peduncle, 1 or 2 times dichotomously branched. Flowers ca. 1 cm in diam, yellow. Capsule tetra-globose, usually with 4 right angles, pinkish or reddish when fresh. Aril orange. Fl. Apr.-Jun., fr. Aug.-Sep..

果枝 Fruiting branchlet

摄影：张金龙 Photo by: Zhang Jinlong

花序 Inflorescence

摄影：陈彬 Photo by: Chen Bin

蒴果 Capsule

摄影：方腾 Photo by: Fang Teng

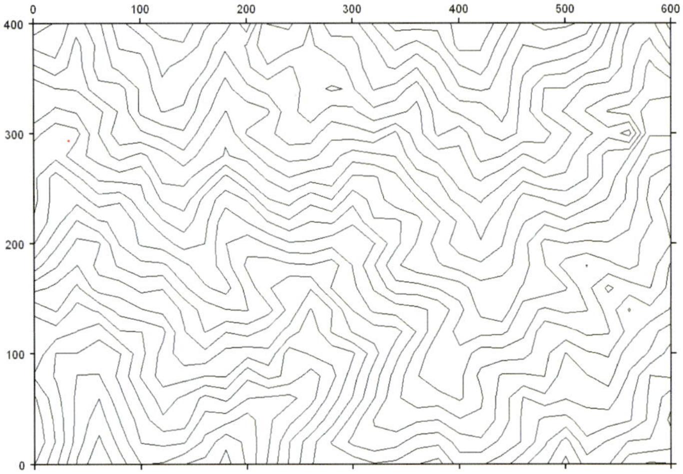

个体分布图 Distribution of Individuals

径级分布表 DBH Class

胸径区间 (Diameter class) (cm)	个体数 (No. of individuals in the plot)	比例 (Proportion)
1～2	0	0.00
2～5	0	0.00
5～10	1	1.00
10～20	0	0.00
20～30	0	0.00
30～60	0	0.00
≥60	0	0.00

83 肉花卫矛 ròuhuāwèimáo | Carnose Euonymus

Euonymus carnosus Hemsl.
卫矛科 | Celastraceae

代码（SpCode）= EUOCAR
个体数（Individual number/24ha）= 18
最大胸径（Max DBH）= 11 cm
重要值排序（Importance value rank）= 103

落叶灌木或小乔木，高达8m。小枝灰绿色到灰褐色，圆柱形，结实，侧脉8～12对。花序1～3次分枝，花数朵到多数。蒴果四角形，具4棱。种子每心室3～6枚，有假种皮。花期5～8月，果期8～11月。

Deciduous shrub to small tree, up to 8 m tall; branches and twigs gray-green to gray-brown, sturdy. Leaf blade thickly papery to leathery; lateral veins 8-12 pairs. Peduncle 1-3 × dichotomously branched with several to many flowers. Capsule 4-angled. Seeds 3-6 per locule, ellipsoid, dark brown, arillate. Fl. May-Aug., fr. Aug.-Nov..

幼苗 Seedling
摄影：方腾 Photo by: Fang Teng

花序 Inflorescence
摄影：方腾 Photo by: Fang Teng

蒴果 Capsule
摄影：杨波 Photo by: Yang Bo

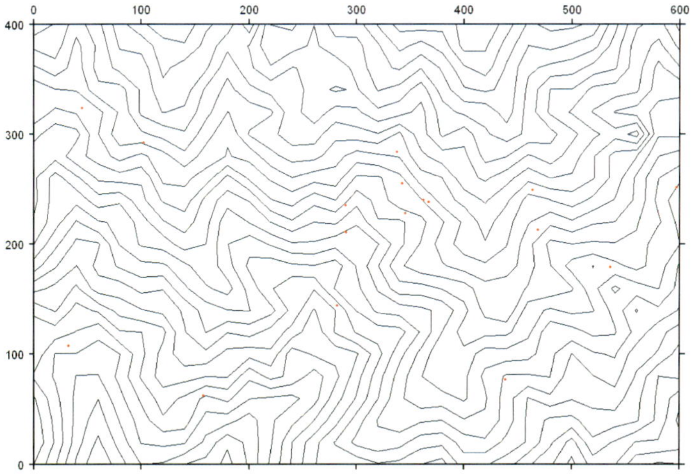

个体分布图 Distribution of Individuals

径级分布表 DBH Class

胸径区间 (Diameter class) (cm)	个体数 (No. of individuals in the plot)	比例 (Proportion)
1～2	7	0.39
2～5	6	0.33
5～10	4	0.22
10～20	1	0.06
20～30	0	0.00
30～60	0	0.00
≥60	0	0.00

84 野鸦椿

yěyāchūn | Common Euscaphis

Euscaphis japonica (Thunb.) Dippel
省沽油科 | Staphyleaceae

代码（SpCode）= EUSJAP
个体数（Individual number/24ha）= 114
最大胸径（Max DBH）= 21.6 cm
重要值排序（Importance value rank）= 67

落叶小乔木或灌木，高2～8m。树皮灰褐色，具纵条纹。小枝及芽红紫色，枝叶揉碎后有臭味。叶厚纸质。果皮软革质，红色；种子近圆形，假种皮黑色，有光泽。花期4～6月，果期8～9月。

Deciduous shrub or small tree, 2-8 m tall; bark grayish brown, striped. Twigs and buds dark purple. Twig and leaves odorous when crushed. Leaflet blades thickly papery. Pericarp softly leathery, red-brown with irregular ribs. Seeds subglobose, arils shining black. Fl. Apr.-Jun., fr. Aug.-Sep..

幼苗 Seedling
摄影：方腾 Photo by: Fang Teng

花序 Inflorescence
摄影：陈彬 Photo by: Chen Bin

果序 Infructescence
摄影：方腾 Photo by: Fang Teng

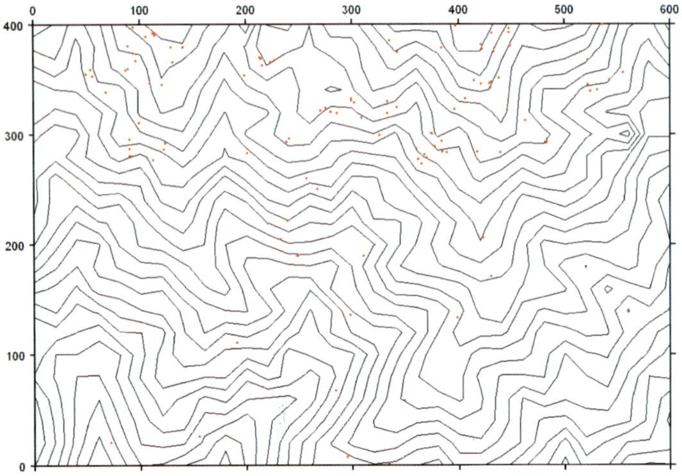

个体分布图 Distribution of Individuals

径级分布表 DBH Class

胸径区间 (Diameter class) (cm)	个体数 (No. of individuals in the plot)	比例 (Proportion)
1～2	22	0.19
2～5	60	0.53
5～10	29	0.25
10～20	2	0.02
20～30	1	0.01
30～60	0	0.00
≥60	0	0.00

85 紫果槭

zǐguǒqì | Purple Fruit Maple

Acer cordatum Pax
槭树科 | Aceraceae

代码（SpCode）= ACECOR
个体数（Individual number/24ha）= 541
最大胸径（Max DBH）= 21.2 cm
重要值排序（Importance value rank）= 41

落叶灌木或小乔木，高达10m。树皮灰色或淡黑灰色，光滑。小枝细瘦，无毛。叶纸质或近革质。伞房花序生于带叶枝顶，花3～5朵。翅果嫩时紫色，成熟时黄褐色，翅张开角度多变。花期4月，果期9月。

Deciduous shrub or small tree, up to 10 m tall. Bark gray, smooth. Branchlets slender, glabrous. Leaf blade papery or semi-leathery. Inflorescence terminal on leafy branchlets, corymbose, 3-5-flowered. Samaras purple when immature, brownish yellow when mature; wings spreading variously. Fl. Apr., fr. Sep..

花枝 Flowering branch
摄影：杨波 Photo by: Yang Bo

幼苗 Seedling
摄影：方腾 Photo by: Fang Teng

翅果 Samara
摄影：方腾 Photo by: Fang Teng

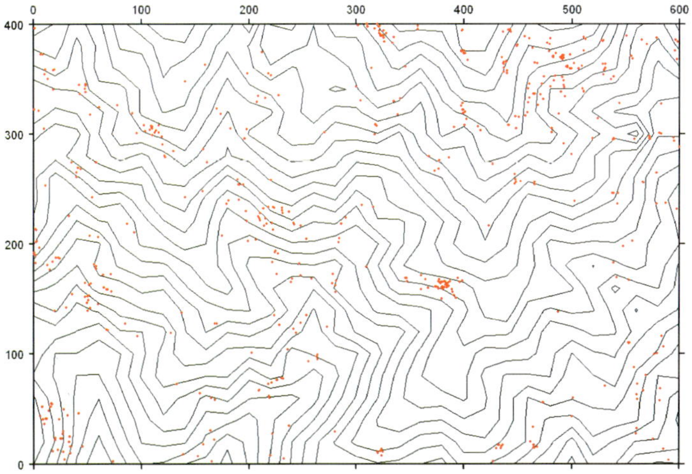

个体分布图 Distribution of Individuals

径级分布表 DBH Class

胸径区间 (Diameter class) (cm)	个体数 (No. of individuals in the plot)	比例 (Proportion)
1～2	228	0.42
2～5	176	0.33
5～10	108	0.20
10～20	28	0.05
20～30	1	0.00
30～60	0	0.00
≥60	0	0.00

86 秀丽槭（橄榄槭） xiùlìqì(gǎnlǎnqì) | Elegant Maple

Acer elegantulum W. P. Fang et P. L. Chiu
槭树科 | Aceraceae

代码（SpCode）= ACEELE
个体数（Individual number/24ha）= 3
最大胸径（Max DBH）= 13.7 cm
重要值排序（Importance value rank）= 134

落叶乔木。高达15m。树皮稍粗糙，深褐色，枝条绿色。叶薄纸质或纸质，叶片宽大于长。圆锥花序生于具2叶小枝顶。翅果嫩时淡紫色，成熟后淡黄色，翅张开近于水平。花期5月，果期9月。

Deciduous tree, up to 15 m tall. Bark dark brown, scabrous; bark on branches green. Leaf blade width somewhat broader than length, thinly papery to papery. Inflorescence paniculate, terminal on a 2-leaved branchlet. Samara glabrous; wings spreading obtusely. Fl. May, fr. Sep..

叶 Leaf
摄影：方腾 Photo by: Fang Teng

花枝 Flowering branch
摄影：陈彬 Photo by: Chen Bin

幼果 Fruit(immature)
摄影：陈彬 Photo by: Chen Bin

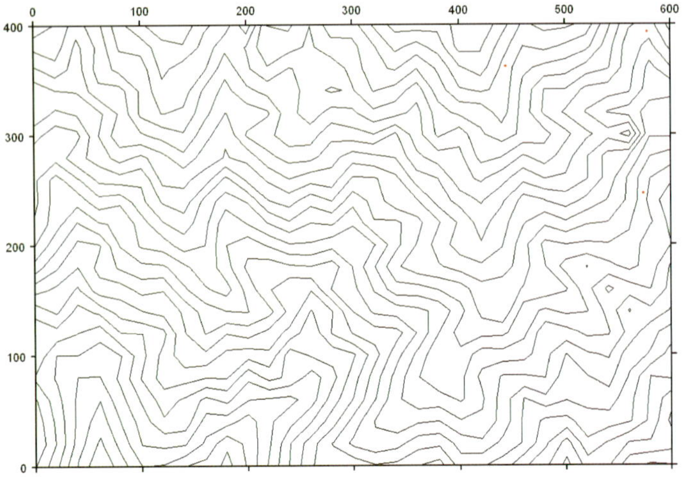

个体分布图 Distribution of Individuals

径级分布表 DBH Class

胸径区间 (Diameter class) (cm)	个体数 (No. of individuals in the plot)	比例 (Proportion)
1～2	0	0.00
2～5	1	0.33
5～10	0	0.00
10～20	2	0.67
20～30	0	0.00
30～60	0	0.00
≥60	0	0.00

87 三峡槭 sānxiáqì | Wilson's Maple

Acer wilsonii Rehd.
槭树科 | Aceraceae

代码（SpCode）= ACEWIL
个体数（Individual number/24ha）= 1
最大胸径（Max DBH）= 18.7 cm
重要值排序（Importance value rank）= 148

落叶乔木，高 10 ~ 15m。树皮深褐色，平滑。小枝细瘦，无毛。当年生嫩枝绿色或淡紫绿色。叶薄纸质，常 3 裂。花杂性，常形成圆锥花序。翅果常成下垂的圆锥果序，黄褐色。花期 4 月，果期 9 月。

Deciduous tree 10-15 m tall. Bark dark brown, smooth. Branchlets slender, glabrous; young twiglets green or viridescence. Leaf blade membranous or papery, usually 3-lobed. Inflorescence andromonoecious, paniculate. Infructescence pendulous, samara yellowish brown. Fl. Apr., fr. Sep..

叶 Leaf
摄影：李晓东 Photo by: Li Xiaodong

果枝 Fruiting branch
摄影：郑小军 Photo by: Zheng Xiaojun

果序 Infructescence
摄影：郑小军 Photo by: Zheng Xiaojun

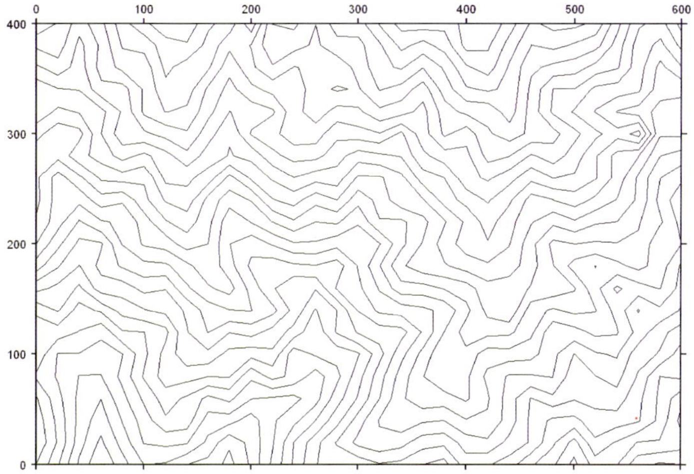

个体分布图 Distribution of Individuals

径级分布表 DBH Class

胸径区间 (Diameter class) (cm)	个体数 (No. of individuals in the plot)	比例 (Proportion)
1～2	0	0.00
2～5	0	0.00
5～10	0	0.00
10～20	1	1.00
20～30	0	0.00
30～60	0	0.00
≥60	0	0.00

88 垂枝泡花树 chuízhīpàohuāshù | Flexuous Meliosma

Meliosma flexuosa Pamp.
清风藤科 | Sabiaceae

代码（SpCode）= MELFLE
个体数（Individual number/24ha）= 12
最大胸径（Max DBH）= 17.6 cm
重要值排序（Importance value rank）= 112

落叶小乔木，高达5m。芽、嫩枝被淡褐色长柔毛，腋芽通常两枚并生。单叶，膜质，侧脉12～18对。圆锥花序顶生，向下弯垂。花白色。核果倒卵形，核极扁斜，网脉突起。花期5～6月，果期7～9月。

Small tree, up to 5 m tall, deciduous. Buds and young branches slightly brownish pilose, usually with 2 buds in leaf axil. Single leaf, blade membranous, lateral veins 12-18 pairs. Panicle terminal, pendulous. Flowers white. Drupes obovoid; endocarp oblique, midrib prominent. Fl. May-Jun., fr. Jul.-Sep..

树干 Trunk
摄影：贾琪 Photo by: Jia Qi

幼苗 Seedling
摄影：方腾 Photo by: Fang Teng

小枝 Branchlet
摄影：贾琪 Photo by: Jia Qi

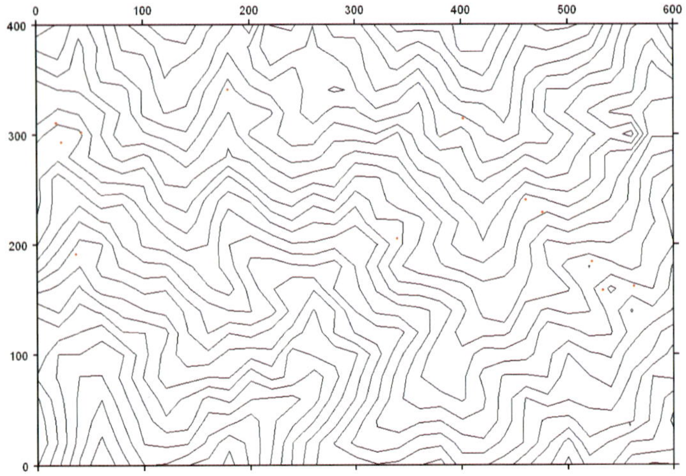

个体分布图 Distribution of Individuals

径级分布表 DBH Class

胸径区间 (Diameter class) (cm)	个体数 (No. of individuals in the plot)	比例 (Proportion)
1～2	0	0.00
2～5	2	0.17
5～10	5	0.42
10～20	5	0.42
20～30	0	0.00
30～60	0	0.00
≥60	0	0.00

89 红枝柴 hóngzhīchái | Oldham's Meliosma

Meliosma oldhamii Maxim.
清风藤科 | Sabiaceae

代码（SpCode）= MELOLD
个体数（Individual number/24ha）= 1146
最大胸径（Max DBH）= 39.7 cm
重要值排序（Importance value rank）= 31

落叶乔木，高达20m。腋芽密被淡褐色柔毛。羽状复叶，小叶薄纸质，侧脉7～8对。圆锥花序顶生，直立。花白色。核果球形，核具明显突起网纹。花期5～6月，果期8～9月。

Deciduous tree, up to 20 m tall. Axillary buds densely light brownish pubescent. Compound leaf, leaflet, thinly papery, lateral veins 7-8 pairs. Panicle terminal, erect. Flowers white. Drupe globose; endocarp convex, with distinct, scattered netlike strips. Fl. May-Jun., fr. Aug.-Sep..

幼苗 Seedling
摄影：金冬梅 Photo by: Jin Dongmei

花序 Inflorescence
摄影：方腾 Photo by: Fang Teng

果枝 Fruiting branch
摄影：方腾 Photo by: Fang Teng

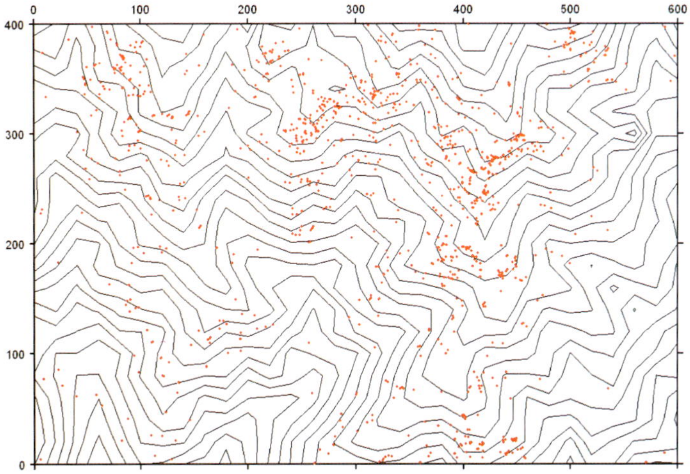

个体分布图 Distribution of Individuals

径级分布表 DBH Class

胸径区间 (Diameter class) (cm)	个体数 (No. of individuals in the plot)	比例 (Proportion)
1～2	303	0.26
2～5	450	0.39
5～10	277	0.24
10～20	79	0.07
20～30	35	0.03
30～60	2	0.00
≥60	0	0.00

90 光叶毛果枳椇 guāngyèmáoguǒzhǐjǔ | Hairyfruit Turnjujube

Hovenia trichocarpa Chun & Tsiang var. *rubusta* (Nakai et Y. Kimura) Y. L. Chen et P. K. Chou
鼠李科 | Rhamnaceae

代码（SpCode）= HOVTRI
个体数（Individual number/24ha）= 4
最大胸径（Max DBH）= 36.6 cm
重要值排序（Importance value rank）= 130

落叶乔木，高达18m。小枝褐色或黑紫色，无毛，有明显皮孔。叶纸质，两面无毛，或仅下面沿脉被疏柔毛。果序轴膨大，有皮孔。核果球形，被密绒毛和长柔毛。种子黑色。花期5～6月，果期8～10月。

Deciduous tree, up to 18 m tall. Branchlets brownish or blackish purple, glabrous, with conspicuous lenticels. Leaf blade papery. Fruiting peduncles and pedicels dilating, lenticelate. Fruit yellow-brown or brown at maturity, subglobose. Seeds black, blackish purple, or brown, densely tomentose. Seeds black. Fl. May-Jun., fr. Aug.-Oct..

幼苗 Seedling
摄影：金冬梅 Photo by: Jin Dongmei

果序 Infructescence
摄影：方腾 Photo by: Fang Teng

果序轴 Fruiting peduncle
摄影：方腾 Photo by: Fang Teng

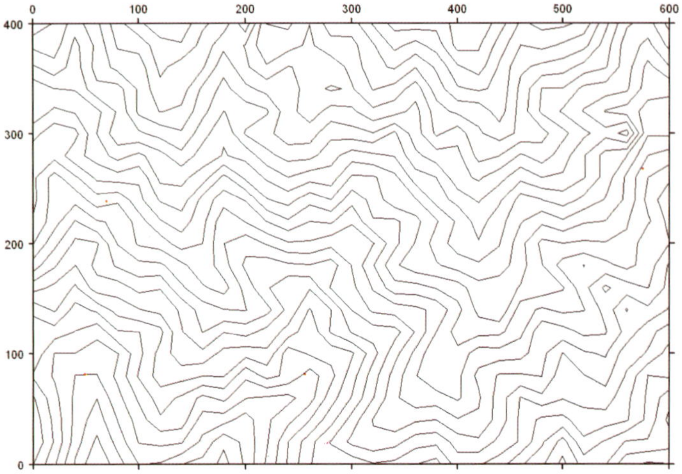

个体分布图 Distribution of Individuals

径级分布表 DBH Class

胸径区间 (Diameter class) (cm)	个体数 (No. of individuals in the plot)	比例 (Proportion)
1～2	0	0.00
2～5	1	0.25
5～10	1	0.25
10～20	0	0.00
20～30	1	0.25
30～60	1	0.25
≥60	0	0.00

91 长叶冻绿　chángyèdònglǜ | Oriental Buckthorn

Rhamnus crenata Sieb. et Zucc.
鼠李科 | Rhamnaceae

代码（SpCode）= RHACRE
个体数（Individual number/24ha）= 54
最大胸径（Max DBH）= 12.5 cm
重要值排序（Importance value rank）= 80

落叶灌木或小乔木，高达7m。枝条被铁锈色毛，后脱落。叶纸质，侧脉8～12条。聚伞花序腋生。核果球形或倒卵球形，幼时绿色或红色，成熟时黑色或紫黑色，核3个。花期5～8月，果期8～10月。

Deciduous shrub or small tree, up to 7 m tall. Young branchlets ferruginous tomentose; old branches reddish, glabrescent. Leaf blade papery, abaxially gray-white tomentose, lateral veins 8-12 pairs. Flowers bisexual, cymes axillary. Drupe green or red when immature, black or purple-black at maturity, globose or obovoid-globose, with 3 one-seeded stones. Fl. May-Aug., fr. Aug.-Oct..

枝叶　Branch and leaves
摄影：陈建华　Photo by: Chen Jianhua

幼果　Fruit(immature)
摄影：陈建华　Photo by: Chen Jianhua

果枝　Fruiting branch
摄影：杨波　Photo by: Yang Bo

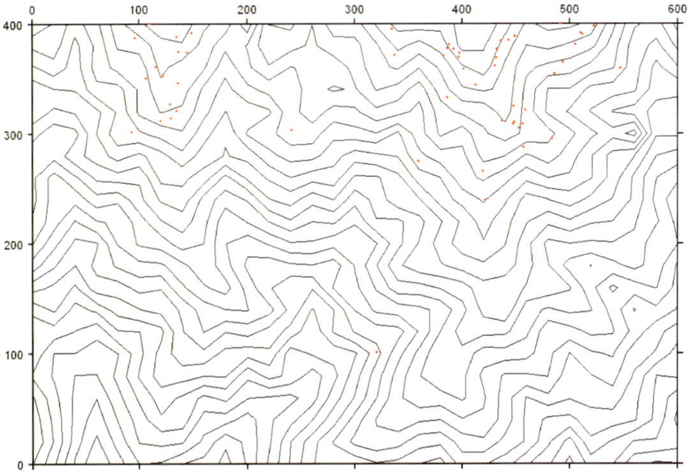

个体分布图 Distribution of Individuals

径级分布表 DBH Class

胸径区间 (Diameter class) (cm)	个体数 (No. of individuals in the plot)	比例 (Proportion)
1～2	17	0.31
2～5	28	0.52
5～10	8	0.15
10～20	1	0.02
20～30	0	0.00
30～60	0	0.00
≥60	0	0.00

92 中华杜英

zhōnghuádùyīng | Chinese Elaeocarpus

Elaeocarpus chinensis (Gardn. et Champ.) Hook. F. ex Benth.
杜英科 | Elaeocarpaceae

代码（SpCode）= ELACHI
个体数（Individual number/24ha）= 22
最大胸径（Max DBH）= 20.6 cm
重要值排序（Importance value rank）= 96

常绿乔木，高达20m。嫩枝有柔毛，老枝秃净。叶薄革质，侧脉4～6对。总状花序生于无叶的去年枝条上。核果椭圆形，长不到1cm。花期5～6月，果期6～9月。

Evergreen tree, up to 20 m tall. Branchlets puberulent, glabrous when old. Leaf blade papery, lateral veins 4-6 pairs. Racemes in axils of fallen leaves. Drupe ellipsoid, shorter than 1 cm. Fl. May-Jun., fr. Jun.-Sep..

幼苗 Seedling
摄影：金冬梅 Photo by: Jin Dongmei

小枝 Branchlet
摄影：方腾 Photo by: Fang Teng

核果 Drupe
摄影：张金龙 Photo by: Zhang Jinlong

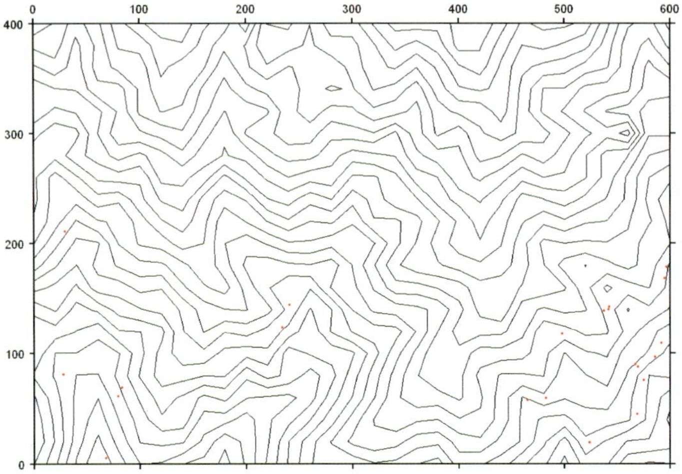

个体分布图 Distribution of Individuals

径级分布表 DBH Class

胸径区间 (Diameter class) (cm)	个体数 (No. of individuals in the plot)	比例 (Proportion)
1～2	6	0.27
2～5	9	0.41
5～10	5	0.23
10～20	1	0.05
20～30	1	0.05
30～60	0	0.00
≥60	0	0.00

93 杜英

dùyīng | Common Elaeocarpus

Elaeocarpus decipiens Hemsl.
杜英科 | Elaeocarpaceae

代码（SpCode）= ELADEC
个体数（Individual number/24ha）= 566
最大胸径（Max DBH）= 27.1 cm
重要值排序（Importance value rank）= 42

常绿乔木，高 5 ~ 15m。嫩枝及顶芽初时被微毛，不久变秃净。叶革质或厚纸质，侧脉 7 ~ 9 对。总状花序生于无叶的去年枝上。核果椭圆形，长 2 ~ 2.5cm。花期 6 ~ 7 月，果期 11 月到翌年 1 月。

Evergreen tree, 5-15 m tall. Branchlets and terminal buds puberulent, glabrescent. Leaf blade leathery or thickly papery, lateral veins 7-9 pairs. Racemes in axils of fallen leaves. Drupe ellipsoid, 2-2.5 cm long. Fl. Jun.-Jul., fr. Nov. to Jan. of next year.

叶 Leaf
摄影：张金龙 Photo by: Zhang Jinlong

果枝 Fruiting branch
摄影：张金龙 Photo by: Zhang Jinlong

核果 Drupe
摄影：陈国科 Photo by: Chen Guoke

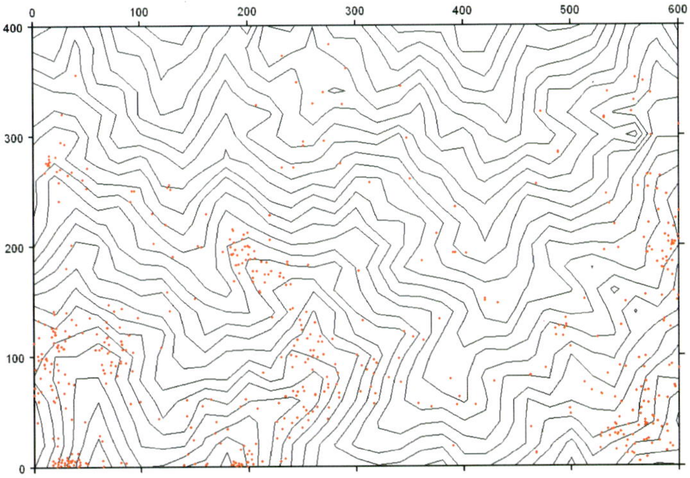

个体分布图 Distribution of Individuals

径级分布表 DBH Class

胸径区间 (Diameter class) (cm)	个体数 (No. of individuals in the plot)	比例 (Proportion)
1～2	158	0.28
2～5	181	0.32
5～10	105	0.19
10～20	99	0.17
20～30	24	0.04
30～60	0	0.00
≥60	0	0.00

94 薯豆

shǔdòu | Japan Elaeocarpus

Elaeocarpus japonicus Sieb. et Zucc.

杜英科 | Elaeocarpaceae

代码（SpCode）= ELAJAP
个体数（Individual number/24ha）= 228
最大胸径（Max DBH）= 45 cm
重要值排序（Importance value rank）= 63

常绿乔木，高达25m。嫩枝秃净无毛，叶芽有发亮绢毛。叶纸质或革质，侧脉5～6对。总状花序生于当年生枝的叶腋内。核果椭圆形，长1～1.3cm。花期4～5月，果期6～7月。

Evergreen tree, up to 25 m tall. Branchlets robust, glabrous or minutely pubescent; buds sericeous. Leaf blade papery or leathery, lateral veins 5-6 pairs. Racemes in axils of leaves at first year twiglet. Drupe shiny, ellipsoid, 1-1.3 cm long. Fl. Apr.-May, fr. Jun.-Jul..

幼苗 Seedling
摄影：方腾 Photo by: Fang Teng

嫩叶和老叶 New leaves and red leaves
摄影：方腾 Photo by: Fang Teng

花序 Inflorescence
摄影：杨波 Photo by: Yang Bo

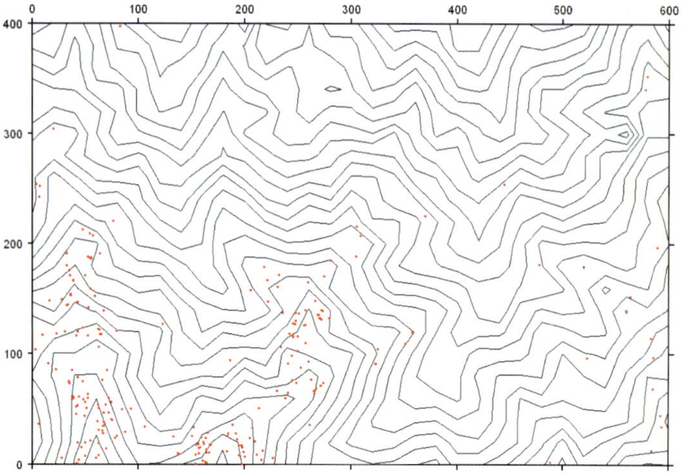

个体分布图 Distribution of Individuals

径级分布表 DBH Class

胸径区间 (Diameter class) (cm)	个体数 (No. of individuals in the plot)	比例 (Proportion)
1～2	46	0.20
2～5	84	0.37
5～10	40	0.18
10～20	38	0.17
20～30	12	0.05
30～60	8	0.04
≥60	0	0.00

95 猴欢喜 hóuhuānxǐ | China Monkeyjoy

Sloanea sinensis (Hance) Hemsl.
杜英科 | Elaeocarpaceae

代码（SpCode）= SLOSIN
个体数（Individual number/24ha）= 43
最大胸径（Max DBH）= 23.1 cm
重要值排序（Importance value rank）= 101

常绿乔木，高达20m。嫩枝无毛。叶柄长1～4cm，叶薄革质，侧脉5～7对。花多朵簇生于枝顶叶腋，花梗长3～6cm。蒴果开裂，针刺长1～1.5cm，内果皮紫红色。花期6～9月，果期翌年5～11月。

Evergreen tree, up to 20 m tall. Branchlets glabrous. Petiole 1-4 cm, leaf blade thinly leathery, lateral veins 5-7 pairs. Flowers fascicled. Pedicel 3-6 cm. Capsule variable valved; prickles 1-1.5 cm; endocarp purple-red. Fl. Jun.-Sep., fr. May-Nov. of next year.

植株 Whole plant
摄影：刘晓娟 Photo by: Liu Xiaojuan

幼苗 Seedling
摄影：方腾 Photo by: Fang Teng

花序 Inflorescence
摄影：方腾 Photo by: Fang Teng

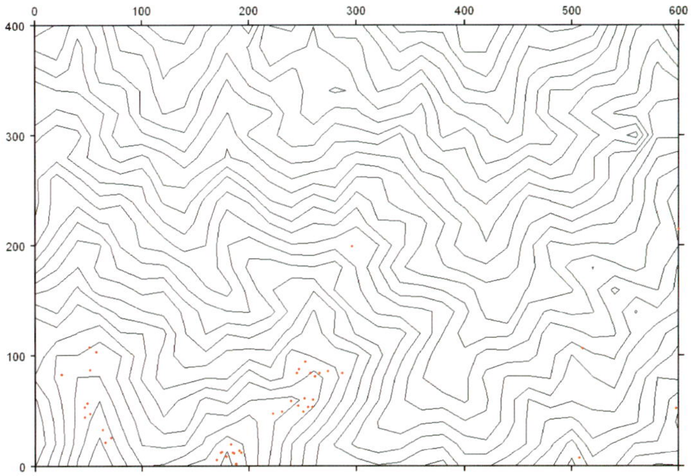

个体分布图 Distribution of Individuals

径级分布表 DBH Class

胸径区间 (Diameter class) (cm)	个体数 (No. of individuals in the plot)	比例 (Proportion)
1～2	12	0.28
2～5	14	0.33
5～10	4	0.09
10～20	9	0.21
20～30	4	0.09
30～60	0	0.00
≥60	0	0.00

96 白毛椴（浆果椴） báimáoduàn(jiāngguǒduàn) | Hunan Linden

Tilia endochrysea Hand.-Mazz.
椴树科 | Tiliaceae

代码（SpCode）= TILEND
个体数（Individual number/24ha）= 1
最大胸径（Max DBH）= 21.1 cm
重要值排序（Importance value rank）= 147

落叶乔木，高达12m。树皮灰色有纵裂。嫩枝鲜绿色，顶芽秃净。叶柄长3～7cm，叶上面无毛，下面被星状绒毛，侧脉5～6对。聚伞花序有花8～32朵。果实球形，5片开裂。花期7～8月，果期10～11月。

Deciduous tree, up to 12 m tall. Bark gray with longitudinal fissures; branchlets bright green when young; buds glabrous. petiole 3-7 cm; leaf blade abaxially stellate tomentose, sometimes glabrescent, adaxially glabrous; lateral veins 5-6 pairs. Cymes with 8-32 flowers. Fruit globose, dehiscent into 5-valves. Fl. Jul.-Aug., fr. Oct.-Nov..

植株 Whole plant
摄影：方腾 Photo by: Fang Teng

嫩叶 New leaves
摄影：方腾 Photo by: Fang Teng

果枝 Fruiting branch
摄影：方腾 Photo by: Fang Teng

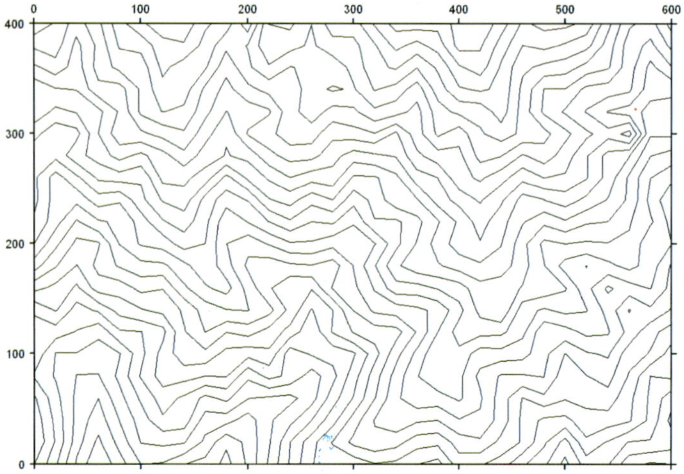

个体分布图 Distribution of Individuals

径级分布表 DBH Class

胸径区间 (Diameter class) (cm)	个体数 (No. of individuals in the plot)	比例 (Proportion)
1～2	0	0.00
2～5	0	0.00
5～10	0	0.00
10～20	0	0.00
20～30	1	1.00
30～60	0	0.00
≥60	0	0.00

97 密花梭罗 mìhuāsuōluó | Denseflower Suoluo

Reevesia pycnantha Ling
梧桐科 | Sterculiaceae

代码（SpCode）= REEPYC
个体数（Individual number/24ha）= 25
最大胸径（Max DBH）= 30.3 cm
重要值排序（Importance value rank）= 93

落叶乔木，高6～10m。小枝灰色，有条纹。叶纸质，全缘或有少数锯齿，侧脉7～8对。聚伞圆锥花序顶生，多花。蒴果椭圆状梨形，顶端截形，密被淡黄褐色短柔毛。种子有翅。花期5～7月，果期9～10月。

Deciduous tree, 6-10 m tall. Branchlets gray, striate. Leaf blade papery, lateral veins 7-8 pairs, margin entire, or sparsely denticulate. Inflorescence thyrsoid, terminal, densely many-flowered. Capsule ellipsoid-pyriform, apex truncate, densely yellowish brown puberulent. Fl. May-Jul., Fr Sep.-Oct..

小枝 Branchlet
摄影：方腾 Photo by: Fang Teng

芽 Bud
摄影：方腾 Photo by: Fang Teng

幼苗 Seedling
摄影：方腾 Photo by: Fang Teng

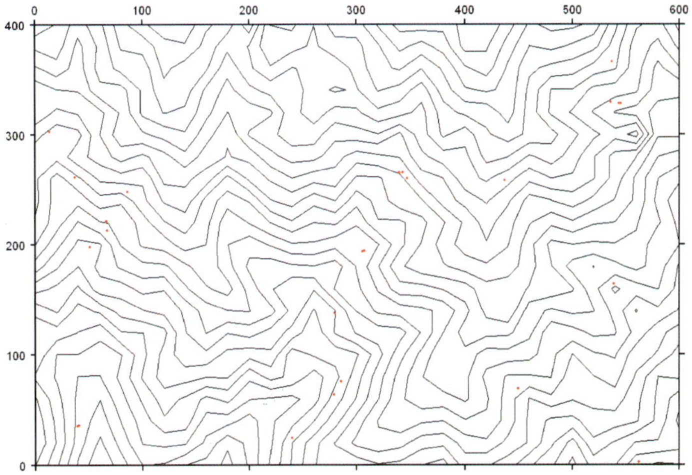

个体分布图 Distribution of Individuals

径级分布表 DBH Class

胸径区间 (Diameter class) (cm)	个体数 (No. of individuals in the plot)	比例 (Proportion)
1～2	1	0.04
2～5	7	0.28
5～10	4	0.16
10～20	6	0.24
20～30	6	0.24
30～60	1	0.04
≥60	0	0.00

98 杨桐（黄瑞木） yángtóng(huángruìmù) | Millett's Adinandra

Adinandra millettii (Hook. et Arn.) Benth. et Hook. f. ex Hance
山茶科 | Theaceae

代码（SpCode）= ADIMIL
个体数（Individual number/24ha）= 664
最大胸径（Max DBH）= 27.4 cm
重要值排序（Importance value rank）= 35

常绿灌木或小乔木，高 2 ~ 10m。树皮灰褐色，当年生小枝灰褐色。叶革质，侧脉 10 ~ 12 对。花单朵腋生，花梗纤细。果圆球形，直径约 1cm，熟时黑色。种子多数。花期 5 ~ 7 月，果期 8 ~ 10 月。

Evergreen shrub or small tree, 2-10 m tall. Bark greyish brown. First year branchlets grayish brown. Leaf blade leathery, lateral veins 10-12 pairs. Flowers axillary, solitary. Pedicel slender. Fruit black when mature, globose, ca. 1 cm in diam, with many seeds. Fl. May-Jul., fr. Aug.-Oct..

花 Flower
摄影：方腾 Photo by: Fang Teng

小枝 Branchlet
摄影：陈彬 Photo by: Chen Bin

浆果 Berry
摄影：方腾 Photo by: Fang Teng

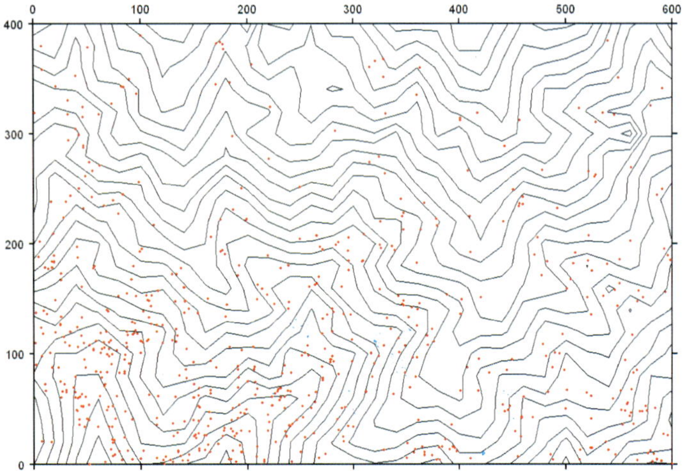

个体分布图 Distribution of Individuals

径级分布表 DBH Class

胸径区间 (Diameter class) (cm)	个体数 (No. of individuals in the plot)	比例 (Proportion)
1～2	168	0.25
2～5	188	0.28
5～10	143	0.22
10～20	151	0.23
20～30	14	0.02
30～60	0	0.00
≥60	0	0.00

99 浙江红山茶

zhèjiānghóngshānchá | Zhejiang Camellia

Camellia chekiangoleosa Hu
山茶科 | Theaceae

代码（SpCode）= CAMCHE
个体数（Individual number/24ha）= 8314
最大胸径（Max DBH）= 16.1 cm
重要值排序（Importance value rank）= 7

常绿灌木或小乔木，高3～6m。多年生枝灰白色，当年生枝红褐色，光滑。叶革质到厚革质，侧脉约8对。花红色，近顶生或腋生，直径7～10cm。蒴果卵球形，种子多数，果皮木质。花期11～12月，果期翌年10月。

Evergreen shrub or small tree, 3-6 m tall. First year branchlets reddish brown, glabrous. Leaf blade leathery to thickly leathery, lateral veins about 8 pairs. Flowers solitary or paired, subterminal or axillary, 7-10 cm in diam. Capsule globose with many seeds and woody pericarp. Fl. Nov.-Dec., fr. Oct. of the following year.

幼苗 Seedling
摄影：金冬梅 Photo by: Jin Dongmei

花 Flower
摄影：陈国科 Photo by: Chen Guoke

蒴果 Capsule
摄影：方腾 Photo by: Fang Teng

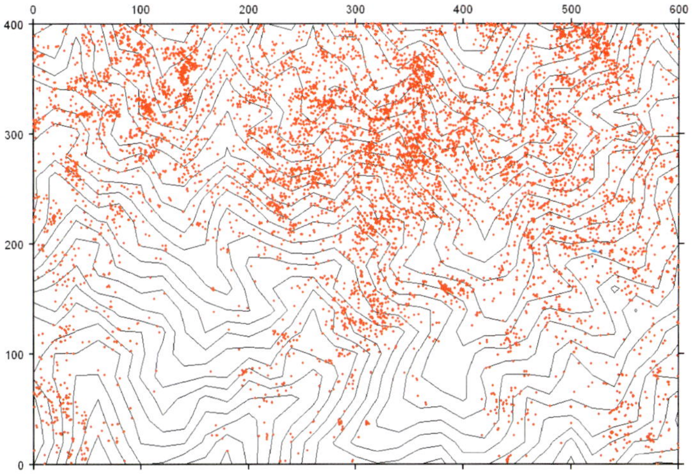

个体分布图 Distribution of Individuals

径级分布表 DBH Class

胸径区间 (Diameter class) (cm)	个体数 (No. of individuals in the plot)	比例 (Proportion)
1～2	5143	0.62
2～5	3015	0.36
5～10	156	0.02
10～20	1	0.00
20～30	0	0.00
30～60	0	0.00
≥60	0	0.00

100 毛花连蕊茶 máohuāliánruǐchá | Hairstalk Tea

Camellia fraterna Hance
山茶科 | Theaceae

代码（SpCode）= CAMFRA
个体数（Individual number/24ha）= 4134
最大胸径（Max DBH）= 21 cm
重要值排序（Importance value rank）= 15

常绿灌木或小乔木，高 1 ~ 5m。嫩枝密生柔毛或长丝毛。叶革质，侧脉 5 ~ 6 对。花腋生，直径 2.5 ~ 4cm，花萼有褐色长丝毛。蒴果圆球形，果壳薄革质，种子 1 个。花期 2 ~ 3 月，果期 9 ~ 10 月。

Evergreen shrub or small tree, 1-5 m tall. First year branchlets densely spreading villous and hirsute. Leaf blade thinly leathery, lateral veins 5-6. Flowers axillary, 2.5-4 cm in diam. Calyx covered with densely yellowish brown villous. Capsule globose, only 1 locule with 1 seed; pericarp thinly leathery. Fl. Feb.-Mar., fr. Sep.-Oct..

幼苗 Seedling
摄影：方腾 Photo by: Fang Teng

小枝 Branchlet
摄影：方腾 Photo by: Fang Teng

花 Flower
摄影：方腾 Photo by: Fang Teng

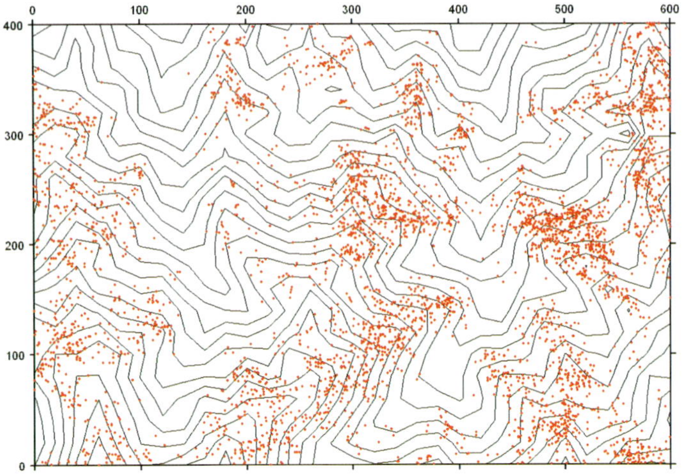

个体分布图 Distribution of Individuals

径级分布表 DBH Class

胸径区间 (Diameter class) (cm)	个体数 (No. of individuals in the plot)	比例 (Proportion)
1 ~ 2	2527	0.61
2 ~ 5	1511	0.37
5 ~ 10	97	0.02
10 ~ 20	0	0.00
20 ~ 30	1	0.00
30 ~ 60	0	0.00
≥60	0	0.00

101 尖连蕊茶 jiānliánruǐchá | Cuspidate Tea

Camellia cuspidata (Kochs) Wright ex Gard.
山茶科 | Theaceae

代码（SpCode）= CAMCUS
个体数（Individual number/24ha）= 29
最大胸径（Max DBH）= 4.1 cm
重要值排序（Importance value rank）= 97

常绿灌木，高达3m。当年生枝稻草黄色，纤细，无毛或幼时有柔毛很快即脱落。叶革质到薄革质，侧脉6～7对。花腋生或近顶生，单生。蒴果圆球形，1心室，1种子。花期12至翌年4月，果期8～10月。

Evergreen shrub up to 3 m tall. First year branchlets straw colored, slender, glabrous or very minutely puberulent when young, soon glabrescent. Leaf blade leathery to thinly leathery, lateral veins 6-7 pairs. Flowers axillary or subterminal, solitary. Capsule globose, only 1 locule with 1 seed. Fl. Dec. to Apr. of next year, fr. Aug.-Oct..

幼苗 Seedling
摄影：陈建华 Photo by: Chen Jianhua

花蕾 Flower bud
摄影：陈建华 Photo by: Chen Jianhua

花 Flower
摄影：陈建华 Photo by: Chen Jianhua

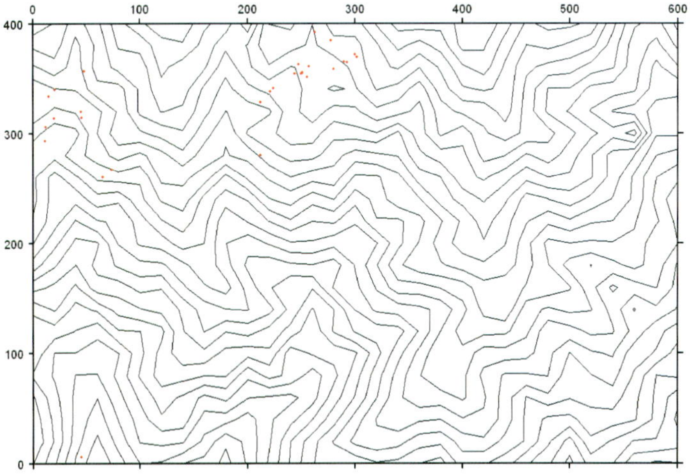

个体分布图 Distribution of Individuals

径级分布表 DBH Class

胸径区间 (Diameter class) (cm)	个体数 (No. of individuals in the plot)	比例 (Proportion)
1～2	24	0.83
2～5	5	0.17
5～10	0	0.00
10～20	0	0.00
20～30	0	0.00
30～60	0	0.00
≥60	0	0.00

102 红淡比

hóngdànbǐ | Japan Cleyera

Cleyera japonica Thunb.

山茶科 | Theaceae

代码（SpCode）= CLEJAP

个体数（Individual number/24ha）= 484

最大胸径（Max DBH）= 34.2 cm

重要值排序（Importance value rank）= 53

常绿灌木或小乔木，高2～10m。全株无毛。树皮灰褐色或灰白色，顶芽1～1.5cm。叶革质。花1～5朵生于叶腋。果实圆球形，成熟时紫黑色，直径8～10mm。种子多数。花期5～6月，果期10～11月。

Evergreen shrub or small tree, 2-10 m tall, glabrous. Bark greyish brown or greyish white; terminal buds 1-1.5 cm. Leaf blade leathery. Flowers axillary, solitary to 5 in a cluster. Fruit purplish black when mature, globose, 8-10 mm in diam with many seeds. Fl. May-Jul., fr. Oct.-Nov..

花 Flowers

摄影：方腾 Photo by: Fang Teng

叶 Leaf

摄影：陈国科 Photo by: Chen Guoke

果枝 Fruiting branch

摄影：方腾 Photo by: Fang Teng

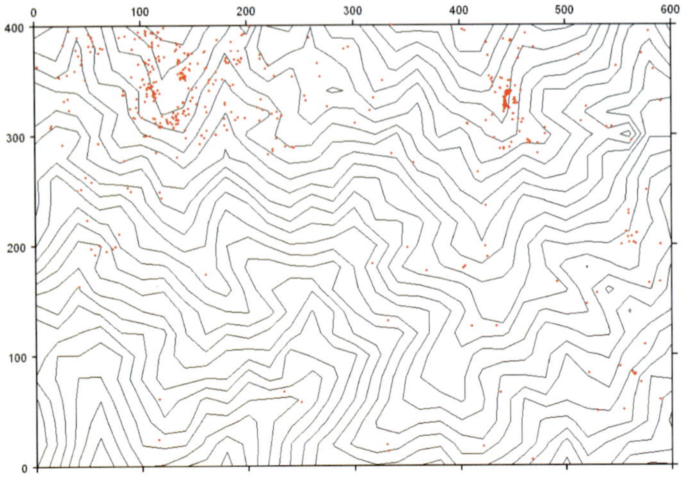

个体分布图 Distribution of Individuals

径级分布表 DBH Class

胸径区间 (Diameter class) (cm)	个体数 (No. of individuals in the plot)	比例 (Proportion)
1～2	244	0.50
2～5	144	0.30
5～10	38	0.08
10～20	53	0.11
20～30	4	0.01
30～60	1	0.00
≥60	0	0.00

103 细枝柃 xìzhīlíng | Slenderbranch Eurya

Eurya loquaiana Dunn
山茶科 | Theaceae

代码（SpCode）= EURLOQ
个体数（Individual number/24ha）= 19
最大胸径（Max DBH）= 4.2 cm
重要值排序（Importance value rank）= 106

常绿灌木或小乔木，高2～10m。一年生枝黄绿色，圆柱形，纤细，微被毛。叶薄革质，侧脉8～20对。花1～4朵簇生于叶腋。果实圆球形，成熟时黑色，直径3～4mm。花期10～12月，果期翌年7～9月。

Evergreen shrub or small tree, 2-10 m tall. First year branchlets yellowish green, terete, slender, puberulent. Leaf blade thinly leathery to papery, lateral veins 8-20 pairs. Flowers axillary, solitary or to 4 in a cluster. Fruit black when mature, globose, 3-4 mm in diam. Fl. Oct.-Dec., fr. Jul.-Sep. of the following year.

小枝 Branchlet
摄影：陈国科 Photo by: Chen Guoke

叶 Leaf
摄影：金冬梅 Photo by: Jin Dongmei

花序 Inflorescence
摄影：张芬耀 Photo by: Zhang Fenyao

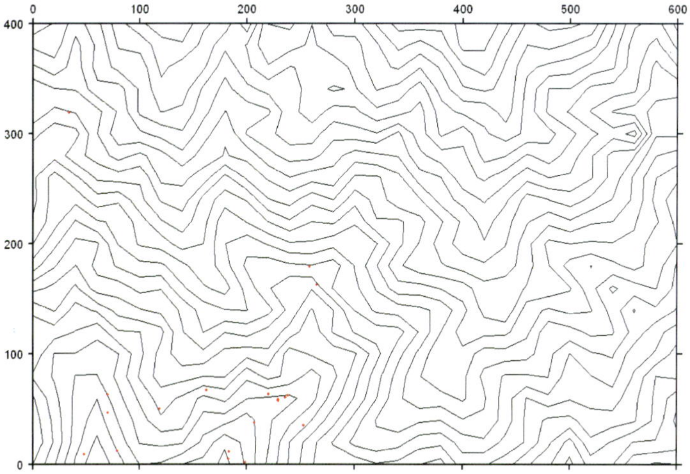

个体分布图 Distribution of Individuals

径级分布表 DBH Class

胸径区间 (Diameter class) (cm)	个体数 (No. of individuals in the plot)	比例 (Proportion)
1～2	11	0.58
2～5	8	0.42
5～10	0	0.00
10～20	0	0.00
20～30	0	0.00
30～60	0	0.00
≥60	0	0.00

104 格药柃 géyàolíng | Muricate Eurya

Eurya muricata Dunn
山茶科 | Theaceae

代码（SpCode）= EURMUR
个体数（Individual number/24ha）= 6610
最大胸径（Max DBH）= 16.9 cm
重要值排序（Importance value rank）= 4

常绿灌木或小乔木，高2～6m，全株无毛。当年小枝生黄绿色。叶革质，侧脉9～11对。花1～5朵簇生于叶腋。果实圆球形，直径4～5mm，成熟时紫黑色。花期9～11月，果期翌年6～9月。

Evergreen shrub or small tree, 2-6 m tall, glabrous. First year branchlets yellowish green. Leaf blade leathery, lateral veins 9-11 pairs. Flowers axillary, solitary or to 5 in a cluster. Fruit purplish black when mature, globose, 4-5 mm in diam. Fl. Sep.-Nov., fr. Jun.-Sep. of the following year.

小枝 Branchlet
摄影：方腾 Photo by: Fang Teng

雄花 Male flower
摄影：方腾 Photo by: Fang Teng

雌花 Female flower
摄影：杨波 Photo by: Yang Bo

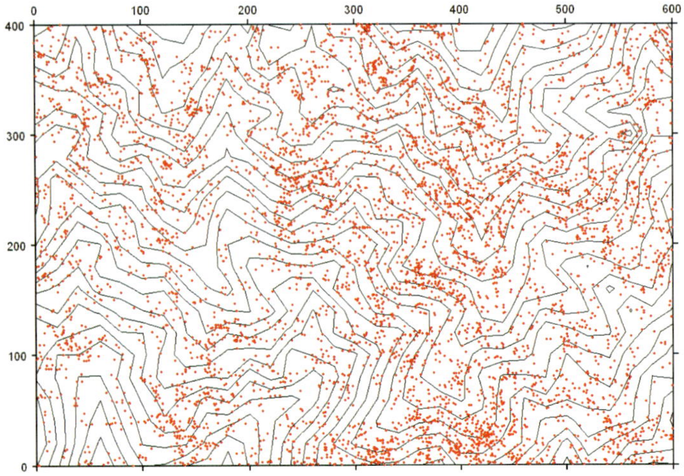

个体分布图 Distribution of Individuals

径级分布表 DBH Class

胸径区间 (Diameter class) (cm)	个体数 (No. of individuals in the plot)	比例 (Proportion)
1～2	2487	0.41
2～5	2772	0.45
5～10	756	0.12
10～20	96	0.02
20～30	0	0.00
30～60	0	0.00
≥60	0	0.00

105 窄基红褐柃 zhǎijīhónghèlíng | Attenuate Eurya

Eurya rubiginosa Hung T. Chang var. *attenuata* H. T. Chang
山茶科 | Theaceae

代码（SpCode）= EURRUB
个体数（Individual number/24ha）= 2769
最大胸径（Max DBH）= 10.2 cm
重要值排序（Importance value rank）= 12

常绿灌木，高2.5～3.5m，全株无毛。当年生枝黄绿色，具明显2棱。叶革质，叶柄长约2mm。花1～3朵簇生于叶腋。果实圆球形或近卵圆形，成熟时紫黑色。花期10～11月，果期翌年4～8月。

Evergreen shrub, 2.5-3.5 m tall, glabrous. Older branches grayish white. First year branchlets yellowish green, 2-ribbed. Petiole ca. 2 mm; leaf blade leathery. Flowers axillary, solitary to 3 in a cluster. Fruit purplish black when mature, globose to ovoid. Fl. Oct.-Nov., fr. Apr.-Aug. of the following year.

花 Flower
摄影：方腾 Photo by: Fang Teng

叶 Leaf
摄影：方腾 Photo by: Fang Teng

小枝 Branchlet
摄影：陈建华 Photo by: Chen Jianhua

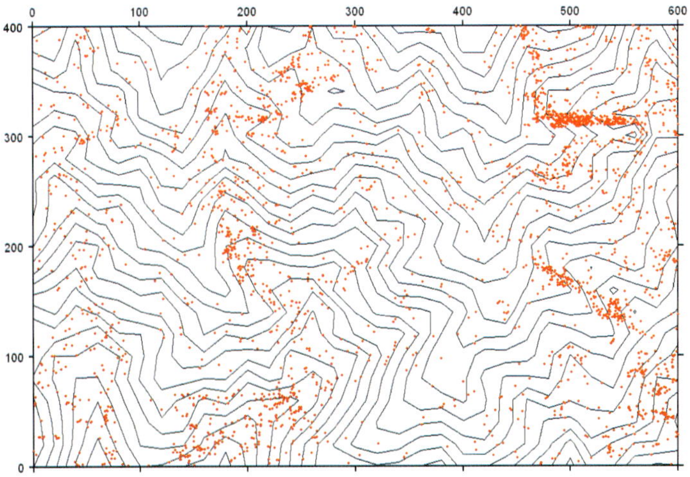

个体分布图 Distribution of Individuals

径级分布表 DBH Class

胸径区间 (Diameter class) (cm)	个体数 (No. of individuals in the plot)	比例 (Proportion)
1～2	1996	0.72
2～5	768	0.28
5～10	5	0.00
10～20	1	0.00
20～30	0	0.00
30～60	0	0.00
≥60	0	0.00

106 木荷

mùhé | Gugertree

Schima superba Gard. et Champ.

山茶科 | Theaceae

代码（SpCode）= SCHSUP

个体数（Individual number/24ha）= 8514

最大胸径（Max DBH）= 83 cm

重要值排序（Importance value rank）= 2

常绿大乔木，高达25m。互生叶，革质或薄革质，椭圆形，边缘有钝齿；叶柄长1～2cm。花生于枝顶叶腋，常4～8朵排成总状花序。蒴果椭球形，直径1.5～2cm。花期6～8月，果期9～12月。

Evergreen tree, up to 25 m tall. Alternate leaves, blade leathery to thinly leathery, elliptic, margin obtusely dentate. Petiole 1-2 cm. Flowers in upper axils of leaves, 4-8 in a raceme. Capsule subglobose, 1.5-2 cm in diam. Fl. Jun.-Aug., fr. Sep.-Dec..

植株 Whole plant

摄影：陈彬 Photo by: Chen Bin

花 Flower

摄影：方腾 Photo by: Fang Teng

果枝 Fruiting branch

摄影：张金龙 Photo by: Zhang Jinlong

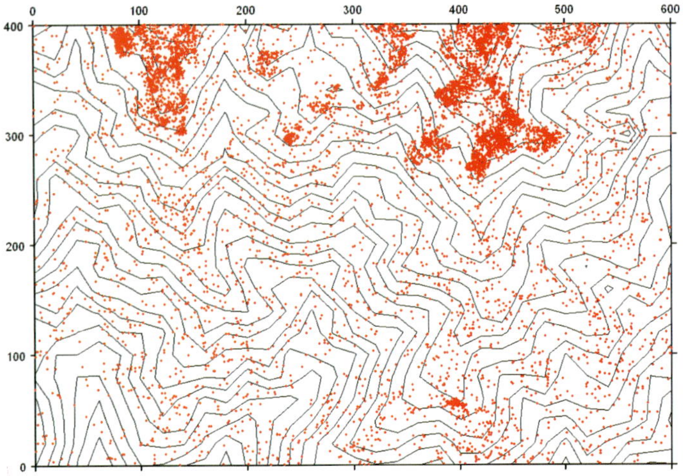

个体分布图 Distribution of Individuals

径级分布表 DBH Class

胸径区间 (Diameter class) (cm)	个体数 (No. of individuals in the plot)	比例 (Proportion)
1～2	2500	0.29
2～5	2208	0.26
5～10	980	0.12
10～20	893	0.10
20～30	1161	0.14
30～60	765	0.09
≥60	7	0.00

107 厚皮香 hòupíxiāng | Ternstroemia

Ternstroemia gymnanthera (Wight et Arn.) Bedd.
山茶科 | Theaceae

代码（SpCode）= TERGYM
个体数（Individual number/24ha）= 3175
最大胸径（Max DBH）= 30.1 cm
重要值排序（Importance value rank）= 9

常绿灌木或小乔木。高1.5～10m。全株无毛。嫩枝浅红褐色或灰褐色。叶革质，通常聚生枝顶，呈假轮生状。花单生无叶的节上。果实圆球形，种子具红色肉质假种皮。花期5～7月，果期9～11月。

Evergreen shrub 1.5-10 m tall, glabrous everywhere. First year branchlets purplish red to reddish brown. Leaves terminal clustered like whorled, blade leathery. Flowers axillary, solitary or several clustered on leafless branchlets. Fruit globose. Aril of seeds fleshy, purplish red when mature. Fl. May-Jul., fr. Sep.-Nov..

叶 Leaf
摄影：方腾 Photo by: Fang Teng

花蕾 Flower bud
摄影：杨波 Photo by: Yang Bo

果枝 Fruiting branch
摄影：谭运洪 Photo by: Tan Yunhong

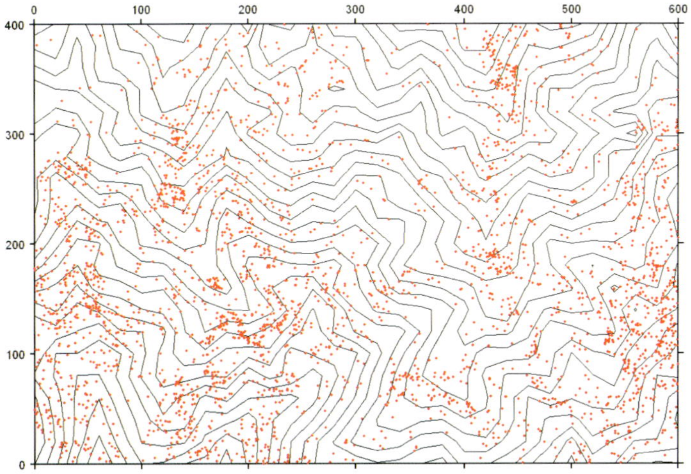

个体分布图 Distribution of Individuals

径级分布表 DBH Class

胸径区间 (Diameter class) (cm)	个体数 (No. of individuals in the plot)	比例 (Proportion)
1～2	1422	0.45
2～5	892	0.28
5～10	349	0.11
10～20	458	0.14
20～30	55	0.02
30～60	1	0.00
≥60	0	0.00

108 小果核果茶（小果石笔木） xiǎoguǒhéguǒchá(xiǎoguǒshíbǐmù) | Smallfruit Pyrenaria

Pyrenaria microcarpa (Dunn) H. Keng
山茶科 | Theaceae

代码（SpCode）= PYRMIC
个体数（Individual number/24ha）= 35
最大胸径（Max DBH）= 14.7 cm
重要值排序（Importance value rank）= 111

常绿灌木或乔木，高1～15m。当年生枝黄褐色，初时有柔毛后变无毛。叶革质，侧脉8～9对，两面皆可见。花细小，白色，直径1.5～2.5cm。蒴果三角球形，三室，每室种子3。花期4～7月，果期8～11月。

Evergreen shrub or small tree, 1-15 m tall. First year branchlets yellowish brown, pubescent, soon glabrescent. Leaf blade leathery, secondary veins 8-9 pairs, visible on both surfaces. Flowers axillary, solitary, 1.5-2.5 cm in diam. Capsule triangular-globose, 3-loculed with 3 seeds per locule. Fl. Apr.-Jul., fr. Aug.-Nov..

幼苗 Seedling
摄影：方腾 Photo by: Fang Teng

小枝 Branchlet
摄影：张金龙 Photo by: Zhang Jinlong

果枝 Fruiting branch
摄影：张金龙 Photo by: Zhang Jinlong

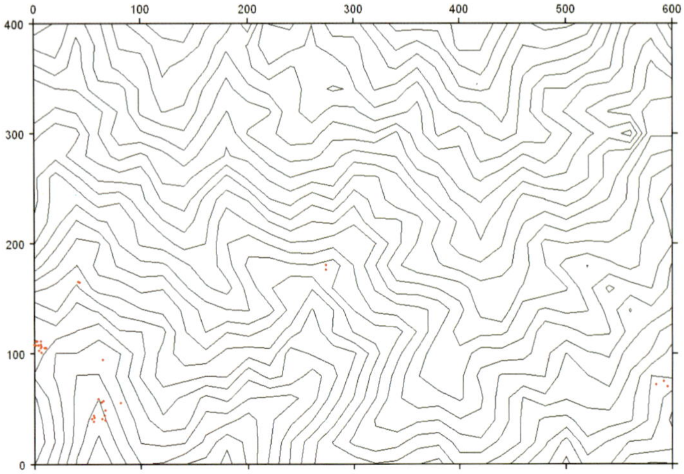

个体分布图 Distribution of Individuals

径级分布表 DBH Class

胸径区间 (Diameter class) (cm)	个体数 (No. of individuals in the plot)	比例 (Proportion)
1～2	12	0.34
2～5	13	0.37
5～10	3	0.09
10～20	7	0.20
20～30	0	0.00
30～60	0	0.00
≥60	0	0.00

109 山桐子 shāntóngzǐ | Manyfruit Idesia

Idesia polycarpa Maxim.
大风子科 | Flacourtiaceae

代码（SpCode）= IDEPOL
个体数（Individual number/24ha）= 28
最大胸径（Max DBH）= 23.6 cm
重要值排序（Importance value rank）= 99

落叶乔木，高8～21m。树皮淡灰色，不裂。小枝黄棕色，有明显皮孔。叶薄革质或厚纸质，边缘有粗齿，齿尖有腺体。圆锥花序顶生下垂，花单性。浆果成熟时紫红色。花期4～5月，果期10～11月。

Deciduous tree, 8-21 m tall; bark grayish, not flaking; branchlets yellowish brown, conspicuously lenticelate. Leaf blade thinly leathery of thickly papery; margin serrate, tip grandular. Panicles terminal, pendulous; flowers unisexual. Berry purple-red or orange-red when mature. Fl. Apr-May, fr. Oct.-Nov..

幼苗 Seedling
摄影：方腾 Photo by: Fang Teng

幼果与叶柄腺体 Immature fruit and gland on petiole
摄影：方腾 Photo by: Fang Teng

果枝 Fruiting branch
摄影：方腾 Photo by: Fang Teng

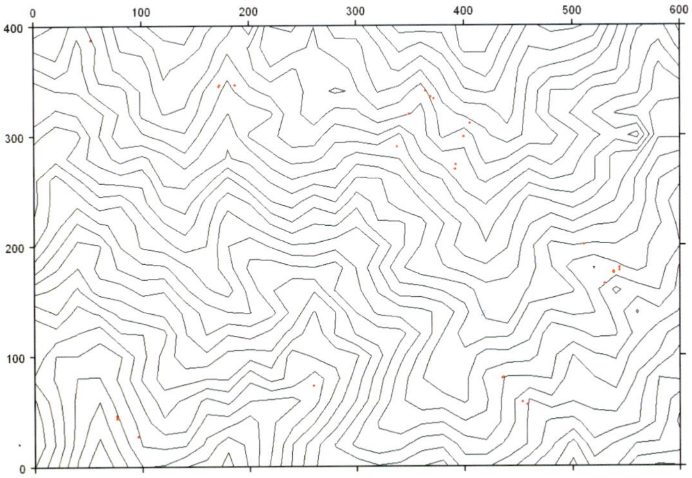

个体分布图 Distribution of Individuals

径级分布表 DBH Class

胸径区间 (Diameter class) (cm)	个体数 (No. of individuals in the plot)	比例 (Proportion)
1～2	7	0.25
2～5	6	0.21
5～10	3	0.11
10～20	9	0.32
20～30	3	0.11
30～60	0	0.00
≥60	0	0.00

110 柞木

zuòmù | Xylosma

Xylosma congesta (Lour.) Merr.

大风子科 | Flacourtiaceae

代码（SpCode）= XYLCON

个体数（Individual number/24ha）= 1

最大胸径（Max DBH）= 18.6 cm

重要值排序（Importance value rank）= 149

常绿大灌木或小乔木，高4～15m，雌雄异株。树皮棕灰色，不规则片状反卷开裂。幼时有枝刺，结果株无刺。叶薄革质。总状花序腋生。浆果黑色，球形，直径4～5mm。花期7～11月，果期8～12月。

Evergreen shrub or small tree, 4-15 m tall, dioecious; bark brown-gray, reflexed flaking; branches spiny when young, unarmed when old. Leaf blade thinly leathery. Inflorescence axillary, racemose. Berry black, globose, 4-5 mm in diam. Fl. Jul.-Nov., fr. Aug.-Dec..

树干 Trunk

摄影：贾琪 Photo by: Jia Qi

叶 Leaf

摄影：贾琪 Photo by: Jia Qi

果枝 Fruiting branch

摄影：陈彬 Photo by: Chen Bin

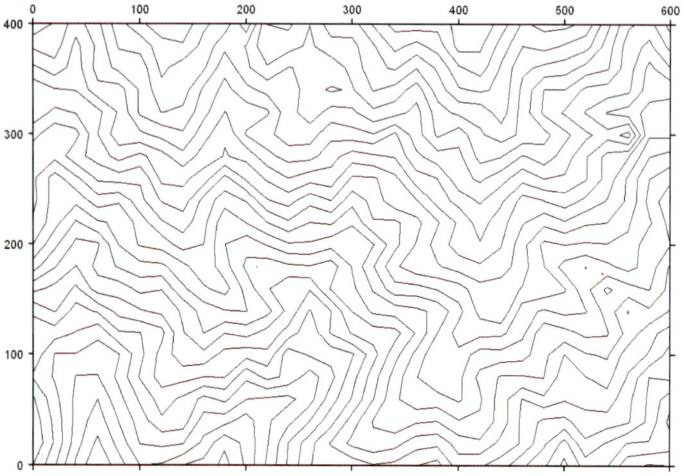

个体分布图 Distribution of Individuals

径级分布表 DBH Class

胸径区间 (Diameter class) (cm)	个体数 (No. of individuals in the plot)	比例 (Proportion)
1～2	0	0.00
2～5	0	0.00
5～10	0	0.00
10～20	1	1.00
20～30	0	0.00
30～60	0	0.00
≥60	0	0.00

111 北江荛花 běijiāngráohuā | Lovely Stringbush

Wikstroemia monnula Hance
瑞香科 | Thymelaeaceae

代码（SpCode）= WIKMON
个体数（Individual number/24ha）= 56
最大胸径（Max DBH）= 3.2 cm
重要值排序（Importance value rank）= 81

落叶灌木，高达2m。枝条黄绿色到暗紫色，有棱角，被柔毛。叶对生或近对生，纸质或坚纸质，侧脉4～5对。总状花序常生于顶端短侧枝顶。核果干燥，卵球形，包被在宿存花萼内。花果期4～8月。

Deciduous shrub, up to 2 m tall. Branches yellowish green to dark purple, ridged, pubescent. Leaves opposite to alternate, blade papery or stiffly so; lateral veins 4-5 pairs. Inflorescences terminal, often on short lateral branches. Drupe dry, ovoid-globose, enclosed by persistent calyx. Fl. and fr. Apr.-Aug..

花序 Inflorescence
摄影：方腾 Photo by: Fang Teng

幼果 Fruit(immature)
摄影：金冬梅 Photo by: Jin Dongmei

果枝 Fruiting branch
摄影：方腾 Photo by: Fang Teng

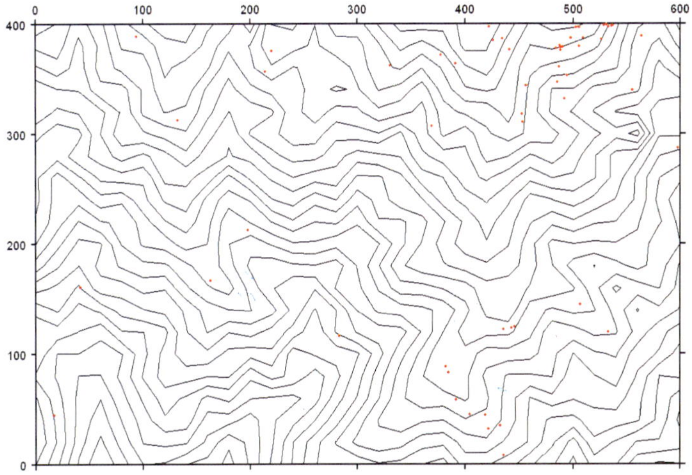

个体分布图 Distribution of Individuals

径级分布表 DBH Class

胸径区间 (Diameter class) (cm)	个体数 (No. of individuals in the plot)	比例 (Proportion)
1～2	49	0.88
2～5	7	0.13
5～10	0	0.00
10～20	0	0.00
20～30	0	0.00
30～60	0	0.00
≥60	0	0.00

112 蓝果树

lánguǒshù | Chinese Tupelo

Nyssa sinensis Oliv.

蓝果树科 | Nyssaceae

代码（SpCode）= NYSSIN
个体数（Individual number/24ha）= 7
最大胸径（Max DBH）= 53.4 cm
重要值排序（Importance value rank）= 123

落叶乔木，高达20m。树皮淡褐色或深灰色，浅裂。当年生枝淡绿色，多年生枝褐色，皮孔显著。叶纸质或薄革质。叶脉6～10对。花序伞形或短总状。果幼时紫绿色，熟时深蓝色。花期4月，果期9月。

Deciduous tree, up to 20 m tall. Bark pale brown to dark gray, shallowly fissured. First year twiglets pea green, older brown, conspicuous lenticelate. Leaf blade papery to subleathery, lateral viens 6-10 pairs. Inflorescences umbels or racemes. Drupe purplish green when immature, bluish at maturity. Fl. Apr., fr. Sep..

幼苗 Seedling
摄影：金冬梅 Photo by: Jin Dongmei

花枝 Flowering branch
摄影：方腾 Photo by: Fang Teng

果枝 Fruiting branch
摄影：方腾 Photo by: Fang Teng

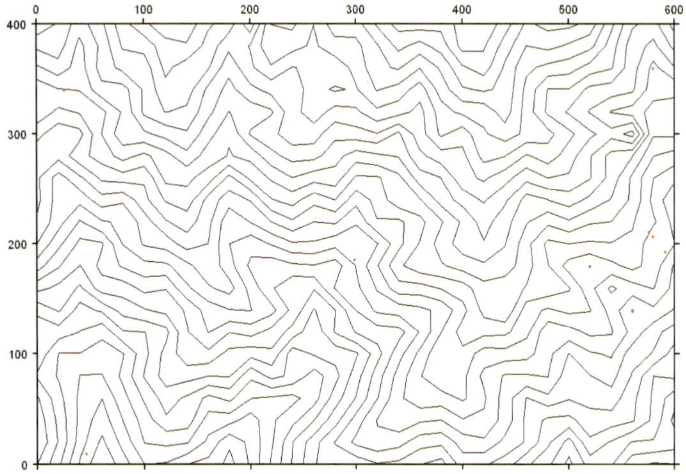

个体分布图 Distribution of Individuals

径级分布表 DBH Class

胸径区间 (Diameter class) (cm)	个体数 (No. of individuals in the plot)	比例 (Proportion)
1～2	0	0.00
2～5	0	0.00
5～10	1	0.14
10～20	1	0.14
20～30	2	0.29
30～60	3	0.43
≥60	0	0.00

113 毛八角枫 máobājiǎofēng | Kurz Alangium

Alangium kurzii Craib
八角枫科 | Alangiaceae

代码（SpCode）= ALAKUR
个体数（Individual number/24ha）= 33
最大胸径（Max DBH）= 18.3 cm
重要值排序（Importance value rank）= 84

落叶小乔木，稀灌木，高5～10m。树皮深褐色，平滑。当年生枝紫绿色，多年生枝深褐色。叶纸质，主脉3～5条，侧脉6～7对，在下面显著。聚伞花序腋生，花5～7朵。核果成熟后近黑色。花期5～6月，果期9月。

Deciduous shrub or small tree, 5-10 m tall. Bark dark brown, smooth. First year twigs purplish green, older dark brown. Leaf blade papery, main viens 3-5, lateral viens 6-7 pairs, conspicuously visible abaxially. Inflorescences 5-7-flowered, axillary. Mature drupe nearly black. Fl. May-Jun., fr. Sep..

幼苗 Seedling
摄影：金冬梅 Photo by: Jin Dongmei

花序 Inflorescence
摄影：方腾 Photo by: Fang Teng

核果 Drupe
摄影：方腾 Photo by: Fang Teng

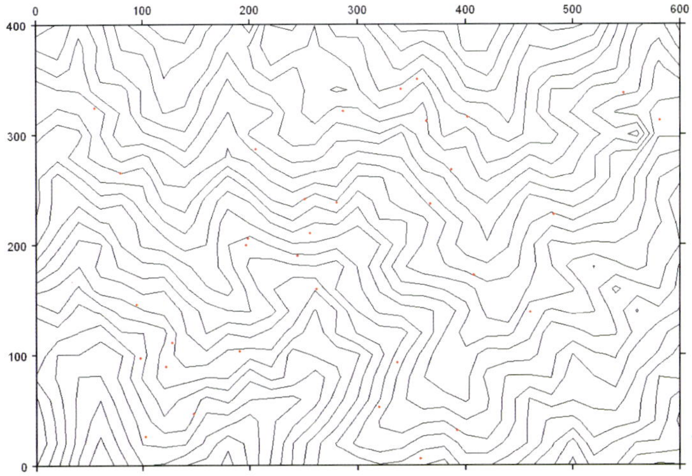

个体分布图 Distribution of Individuals

径级分布表 DBH Class

胸径区间 (Diameter class) (cm)	个体数 (No. of individuals in the plot)	比例 (Proportion)
1～2	7	0.21
2～5	10	0.30
5～10	12	0.36
10～20	4	0.12
20～30	0	0.00
30～60	0	0.00
≥60	0	0.00

114 赤楠

chìnán | Boxleaf Syzygium

Syzygium buxifolium Hook. et Arn.

桃金娘科 | Myrtaceae

代码（SpCode）= SYZBUX

个体数（Individual number/24ha）= 3427

最大胸径（Max DBH）= 18.9 cm

重要值排序（Importance value rank）= 16

常绿灌木或小乔木。嫩枝有棱，干后黑褐色。叶对生或三叶轮生，叶片革质，侧脉多而密，间距1～1.5mm，在下面稍突起。聚伞花序顶生，有花数朵。果实红色到紫黑色，直径5～7mm。花期6～8月，果期10～12月。

Evergreen shrub or small tree. Twigs ridged, blackish brown when dry. Leaves opposite or ternate; leaf blade leathery, secondary veins numerous, 1-1.5 mm apart, abaxially slightly raised. Inflorescences terminal, cymes, several-flowered. Fruit red to purplish black, 5-7 mm in diam. Fl. Jun.-Aug., fr. Oct.-Dec..

幼苗 Seedling

摄影：方腾 Photo by: Fang Teng

花枝 Flowering branch

摄影：陈建华 Photo by: Chen Jianhua

浆果 Berry

摄影：陈彬 Photo by: Chen Bin

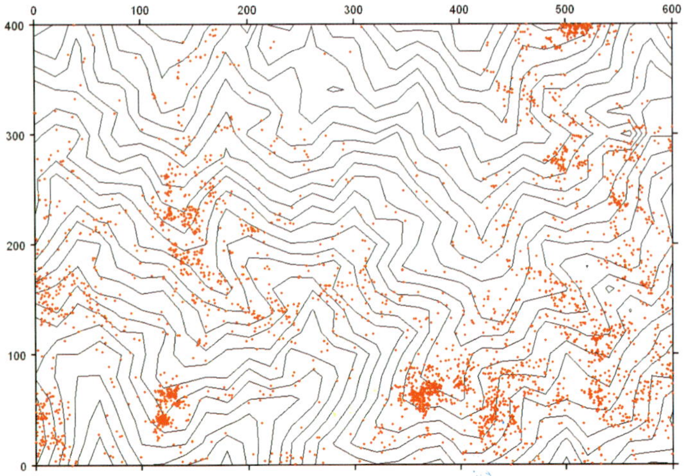

个体分布图 Distribution of Individuals

径级分布表 DBH Class

胸径区间 (Diameter class) (cm)	个体数 (No. of individuals in the plot)	比例 (Proportion)
1～2	1907	0.56
2～5	1080	0.32
5～10	389	0.11
10～20	51	0.01
20～30	0	0.00
30～60	1	0.00
≥60	0	0.00

115 楤木 sǒngmù | China Aralia

Aralia chinensis L.
五加科 | Araliaceae

代码（SpCode）= ARACHI
个体数（Individual number/24ha）= 15
最大胸径（Max DBH）= 3 cm
重要值排序（Importance value rank）= 119

落叶灌木或乔木，高2～5m。树皮灰色。小枝疏生细刺。二回或三回羽状复叶，小叶纸质至薄革质。顶生大型由小伞形花序组成的圆锥花序。果实球形到卵球形，黑色。花期7～9月，果期9～12月。

Deciduous shrub or small tree, 2-5 m tall. Bark greyish. Branches prickly. Leaves 2- or 3- pinnately compound, leaflet blade papery to semi-leathery. Inflorescence a terminal panicle of umbels. Fruit globose to ovoid-globose, blackish. Fl. Jul.-Sep., fr. Sep.-Dec..

植株 Whole plant
摄影：张金龙 Photo by: Zhang Jinlong

幼苗 Seedling
摄影：金冬梅 Photo by: Jin Dongmei

叶 Leaf
摄影：金冬梅 Photo by: Jin Dongmei

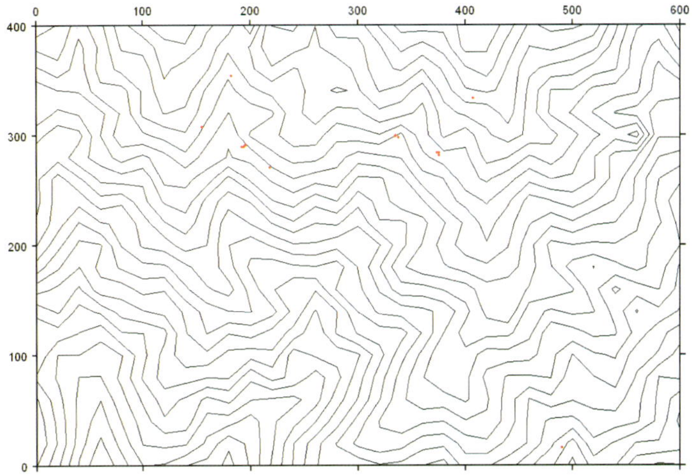

个体分布图 Distribution of Individuals

径级分布表 DBH Class

胸径区间 (Diameter class) (cm)	个体数 (No. of individuals in the plot)	比例 (Proportion)
1～2	11	0.73
2～5	4	0.27
5～10	0	0.00
10～20	0	0.00
20～30	0	0.00
30～60	0	0.00
≥60	0	0.00

116 树参

shùshēn | Tree Renshen

Dendropanax dentiger (Harms) Merr.

五加科 | Araliaceae

代码（SpCode）= DENDEN

个体数（Individual number/24ha）= 4

最大胸径（Max DBH）= 9.1 cm

重要值排序（Importance value rank）= 135

常绿小乔木或灌木，高2～8m。叶形多变。叶片厚纸质或革质。基脉三出，侧脉4～6对。伞形花序顶生，单生或2～5个组成复伞形花序。果实椭圆形到近球形。花期6月，8～9月，果期7月，10～11月。

Evergreen shrub or small tree, 2-10 m tall. Leaf blade sometimes dimorphic, thickly papery to leathery, palmately ternate veined, secondary veins 4-6 pairs. Inflorescence terminal; umbels solitary or 2-5. Fruit ellipsoid to subglobose. Fl. Jun., Aug.-Sep., fr. Jul., Oct.-Nov..

幼苗 Seedling

摄影：方腾 Photo by: Fang Teng

花序 Inflorescence

摄影：方腾 Photo by: Fang Teng

果枝 Fruiting branch

摄影：杨波 Photo by: Yang Bo

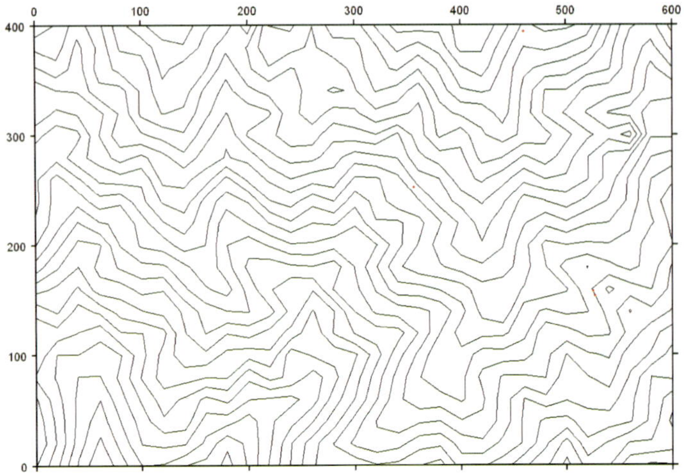

个体分布图 Distribution of Individuals

径级分布表 DBH Class

胸径区间 (Diameter class) (cm)	个体数 (No. of individuals in the plot)	比例 (Proportion)
1～2	0	0.00
2～5	2	0.50
5～10	2	0.50
10～20	0	0.00
20～30	0	0.00
30～60	0	0.00
≥60	0	0.00

117 四照花 sìzhàohuā | Four-involucre

Cornus kousa Buerger ex Miq. subsp. *chinensis* (Osborn) Q. Y. Xiang
山茱萸科 | Cornaceae

代码（SpCode）= CORKOU
个体数（Individual number/24ha）= 1
最大胸径（Max DBH）= 4.6 cm
重要值排序（Importance value rank）= 153

落叶小乔木，高3～10m。叶对生，纸质或厚纸质，背面粉绿色，侧脉4～5对。头状花序球形，由20～40朵花聚集而成，总苞片4，白色。果序球形，成熟时红色。花期5～6月，果期9～10月。

Deciduous shrub or small tree, 3-10 m tall. Leaf opposite, blade papery to thickly papery, abaxially powder green, lateral veins 4-5 pairs. Capitate cymes globose, 20-40-flowered; involucral bract 4, white. Compound fruit red at maturity, globose. Fl. May-Jun., fr. Sep.-Oct..

花序 Inflorescence
摄影：陈彬 Photo by: Chen Bin

枝叶 Branch and leaves
摄影：陈彬 Photo by: Chen Bin

聚花果 Multiple fruit
摄影：陈彬 Photo by: Chen Bin

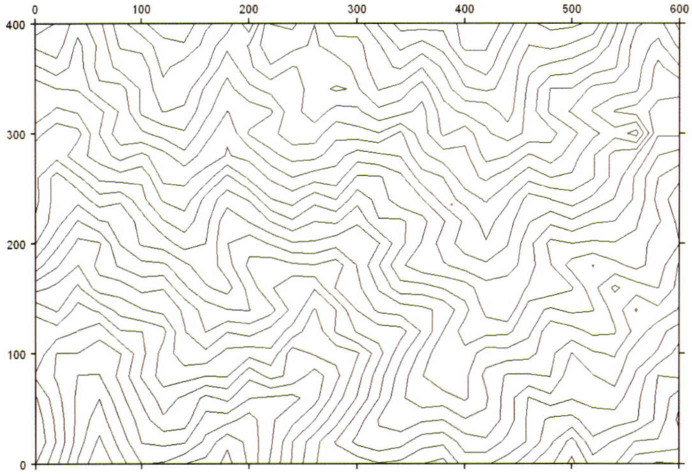

个体分布图 Distribution of Individuals

径级分布表 DBH Class

胸径区间 (Diameter class) (cm)	个体数 (No. of individuals in the plot)	比例 (Proportion)
1～2	0	0.00
2～5	1	1.00
5～10	0	0.00
10～20	0	0.00
20～30	0	0.00
30～60	0	0.00
≥60	0	0.00

118 毛果珍珠花

máoguǒzhēnzhūhuā | Common Lyonia

Lyonia ovalifolia (Wall.) Drude var. *hebecarpa* (Franch. ex Forb. & Hemsl.) Chun

杜鹃花科 | Ericaceae

代码（SpCode）= LYOOVA

个体数（Individual number/24ha）= 419

最大胸径（Max DBH）= 12 cm

重要值排序（Importance value rank）= 62

落叶灌木或小乔木，高1～4m。叶革质。总状花序生于叶腋，近基部常有2～3枚叶状苞片，花冠圆筒状。蒴果球形或卵形，具加厚的缝合线，密被毛。花期5～6月，果期7～9月。

Deciduous shrub or small tree, 1-4 m tall. Leaf blade ovate to elliptic-ovate, leathery. Inflorescence a raceme, lower usually with 2-3 leaflike bracts. Corolla tubular. Capsule globose or ovoid, with thickened sutures, densely pubescent. Fl. May-Jun., fr. Jul.-Sep..

叶 Leaf

摄影：陈建华 Photo by: Chen Jianhua

小枝 Branchlet

摄影：陈建华 Photo by: Chen Jianhua

花序 Inflorescence

摄影：方腾 Photo by: Fang Teng

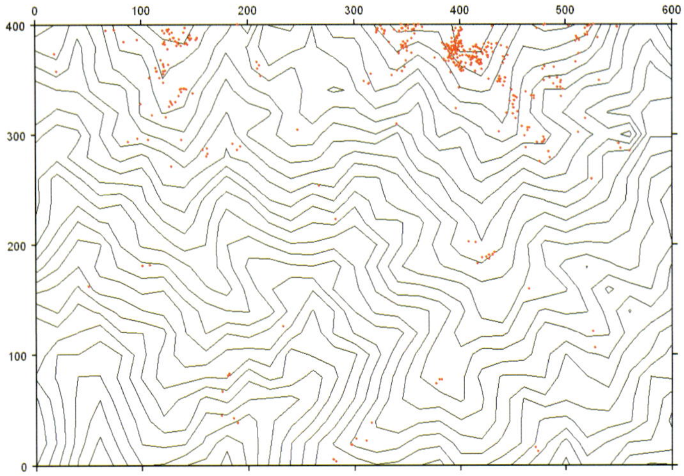

个体分布图 Distribution of Individuals

径级分布表 DBH Class

胸径区间 (Diameter class) (cm)	个体数 (No. of individuals in the plot)	比例 (Proportion)
1～2	137	0.33
2～5	239	0.57
5～10	36	0.09
10～20	6	0.01
20～30	0	0.00
30～60	1	0.00
≥60	0	0.00

119 美丽马醉木 měilìmǎzuìmù | Himalayas Pieris

Pieris formosa (Wall.) D. Don
杜鹃花科 | Ericaceae

代码（SpCode）= PIEFOR
个体数（Individual number/24ha）= 1259
最大胸径（Max DBH）= 42 cm
重要值排序（Importance value rank）= 40

常绿灌木或小乔木，高2～4m。叶革质，基部楔形至钝圆形，叶基部以上有锯齿。总状花序簇生于枝顶的叶腋，或有时为顶生圆锥花序。花冠白色，坛状。蒴果卵圆形，种子纺锤形。花期5～6月，果期7～9月。

Evergreen shrub or small tree, 2-4 m tall. Leaf blade leathery, base cuneate to obtuse, margin serrate from base. Inflorescences paniculate or racemose, terminal axilary. Corolla tubular-urceolate or urceolate. Capsule ovoid. Seeds spindle-shaped. Fl. May-Jun., fr. Jul.-Sep..

幼苗 Seedling
摄影：金冬梅 Photo by: Jin Dongmei

花序 Inflorescence
摄影：方腾 Photo by: Fang Teng

果枝 Fruiting branch
摄影：张金龙 Photo by: Zhang Jinlong

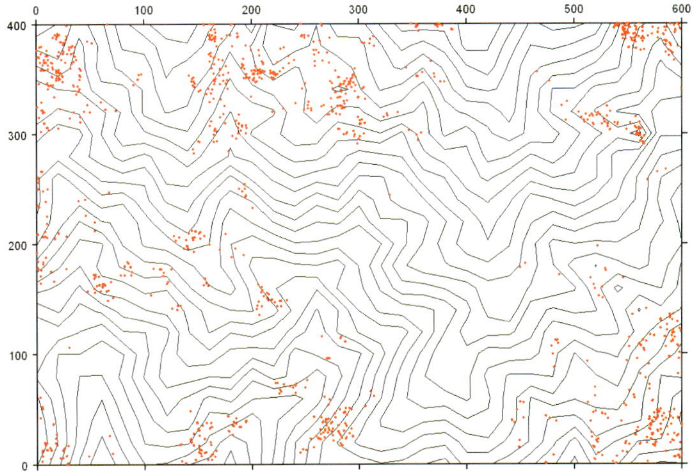

个体分布图 Distribution of Individuals

径级分布表 DBH Class

胸径区间 (Diameter class) (cm)	个体数 (No. of individuals in the plot)	比例 (Proportion)
1～2	306	0.24
2～5	651	0.52
5～10	286	0.23
10～20	14	0.01
20～30	0	0.00
30～60	2	0.00
≥60	0	0.00

120 马醉木

mǎzuìmù | Japan Pieris

Pieris japonica (Thunb.) D. Don ex G. Don
杜鹃花科 | Ericaceae

代码（SpCode）= PIEJAP
个体数（Individual number/24ha）= 2
最大胸径（Max DBH）= 7.1 cm
重要值排序（Importance value rank）= 142

常绿灌木或小乔木，高达4m。树皮棕褐色。叶片革质，基部狭楔形，自中部以上有锯齿。总状花序或圆锥花序顶生或腋生。蒴果近于扁球形。花期2～5月，果期7～9月。

Evergreen shrub or small tree, up to 4 m tall. Leaf blade leathery, base cuneate-attenuate, margin serrate from middle. Inflorescences paniculate or racemose, terminal or axillary. Capsule depressed-globose. Fl. Feb.-May, fr. Jul.-Sep..

叶 Leaf
摄影：张金龙 Photo by: Zhang Jinlong

小枝 Branchlet
摄影：陈国科 Photo by: Chen Guoke

花序 Inflorescence
摄影：刘晓娟 Photo by: Liu Xiaojuan

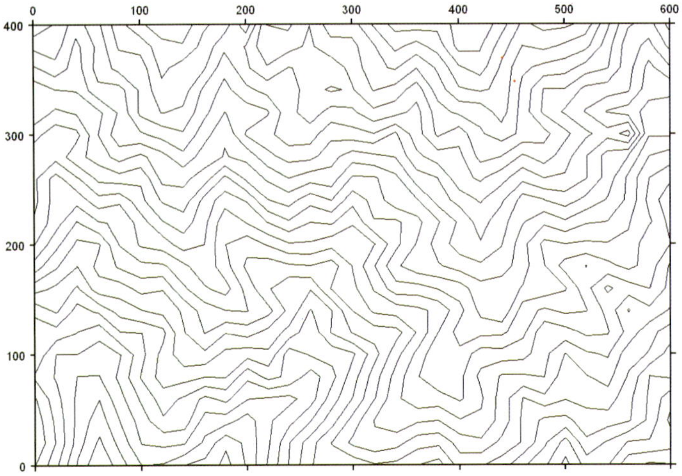

个体分布图 Distribution of Individuals

径级分布表 DBH Class

胸径区间 (Diameter class) (cm)	个体数 (No. of individuals in the plot)	比例 (Proportion)
1～2	0	0.00
2～5	1	0.50
5～10	1	0.50
10～20	0	0.00
20～30	0	0.00
30～60	0	0.00
≥60	0	0.00

121 鹿角杜鹃

jǐjiǎodùjuān | Deerhorn Azalea

Rhododendron latoucheae Franch.

杜鹃花科 | Ericaceae

代码（SpCode）= RHOLAT

个体数（Individual number/24ha）= 2805

最大胸径（Max DBH）= 17.8 cm

重要值排序（Importance value rank）= 13

常绿灌木或小乔木，高2～7m。叶集生枝顶，近于轮生，革质，下面淡灰白色。花单生枝顶叶腋，枝端具花1～4朵。花冠直径约5cm。蒴果圆柱形，花柱宿存。花期3～5月，果期7～10月。

Evergreen shrub or small tree, 2-7 m tall. Leaves terminal clustered at twiglets, pseudoverticillate, leathery, abaxially greyish-white. Inflorescence axillary, subapical 1-4-flowered, corolla pink to white, ca. 5 cm in diam. Capsules cylindric, style persistent. Fl. Mar.-May, fr. Jul.-Oct..

幼苗 Seedling

摄影：方腾 Photo by: Fang Teng

新芽 Sprout

摄影：方腾 Photo by: Fang Teng

花 Flower

摄影：陈建华 Photo by: Chen Jianhua

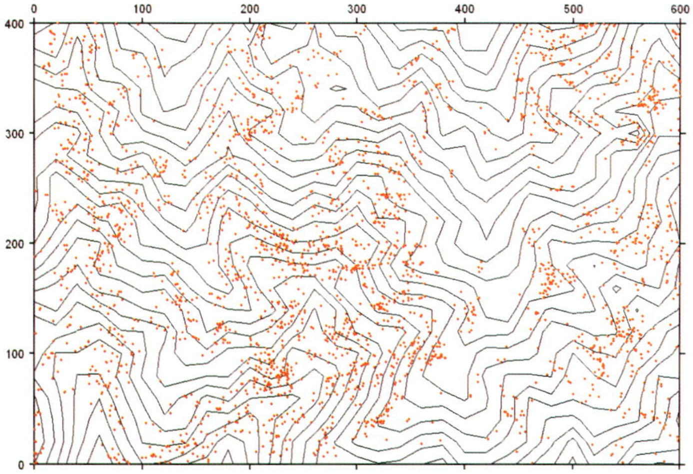

个体分布图 Distribution of Individuals

径级分布表 DBH Class

胸径区间 (Diameter class) (cm)	个体数 (No. of individuals in the plot)	比例 (Proportion)
1～2	456	0.16
2～5	1095	0.39
5～10	1103	0.39
10～20	151	0.05
20～30	0	0.00
30～60	1	0.00
≥60	0	0.00

122 丁香杜鹃（满山红） dīngxiāngdùjuān(mǎnshānhóng) | Clove Azalea

Rhododendron farrerae Tate ex Sweet

杜鹃花科 | Ericaceae

代码（SpCode）= RHOFAR

个体数（Individual number/24ha）= 1741

最大胸径（Max DBH）= 20 cm

重要值排序（Importance value rank）= 50

落叶灌木。高1～4m。枝轮生，幼时被淡黄棕色柔毛，成长时无毛。叶厚纸质或近革质，常2～3集生枝顶。花通常2朵顶生，先花后叶。蒴果密被亮棕褐色长柔毛。花期4～5月，果期6～11月。

Deciduous shrub, 1-4 m tall; branches verticillate, softly yellowish-brown hairy, glabrescent. Leaf blade papery to almost leathery, always 2-3 terminal clustered. Inflorescence usually 2-flowered; flowers appearing before the leaves expansion. Capsule ellipsoid-ovoid, densely shiny-brown-villous. Fl. Apr.-May, fr. Jun.-Nov..

叶 Leaf

摄影：陈彬 Photo by: Chen Bin

花 Flower

摄影：陈彬 Photo by: Chen Bin

蒴果 Capsule

摄影：杨波 Photo by: Yang Bo

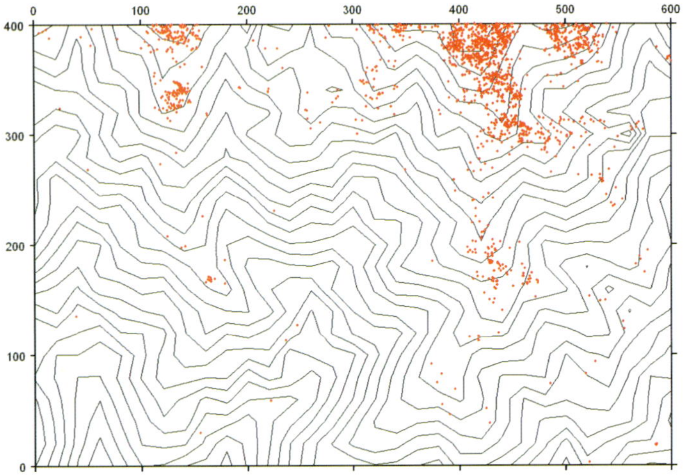

个体分布图 Distribution of Individuals

径级分布表 DBH Class

胸径区间 (Diameter class) (cm)	个体数 (No. of individuals in the plot)	比例 (Proportion)
1～2	849	0.49
2～5	855	0.49
5～10	36	0.02
10～20	1	0.00
20～30	0	0.00
30～60	0	0.00
≥60	0	0.00

123 马银花 mǎyínhuā | Azaleastra

Rhododendron ovatum (Lindl.) Planch. ex Maxim.
杜鹃花科 | Ericaceae

代码（SpCode）= RHOOVA
个体数（Individual number/24ha）= 10791
最大胸径（Max DBH）= 16.6 cm
重要值排序（Importance value rank）= 3

常绿灌木，高2～6m。小枝灰褐色，疏被具柄腺体和短柔毛。叶柄具狭翅，叶革质。花单生枝顶叶腋。蒴果密被灰褐色短柔毛和疏腺体。花期4～5月，果期7～10月。

Evergreen shrub, 2-6 m tall; young shoots sparsely glandular-hairy and pubescent. Petiole narrowly winged, leaf blade leathery; flowers subapical axillary. Capsule densely gray-brown pubescent, also sparsely glandular. Fl. Apr.-May, fr. Jul.-Oct..

幼苗 Seedling
摄影：金冬梅 Photo by: Jin Dongmei

叶 Leaf
摄影：金冬梅 Photo by: Jin Dongmei

花 Flower
摄影：方腾 Photo by: Fang Teng

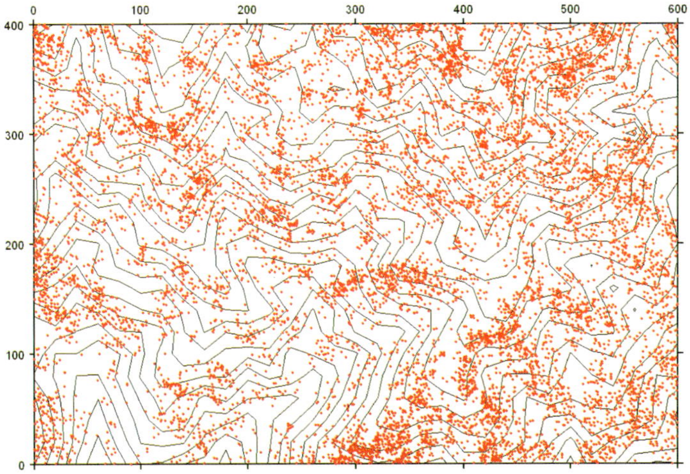

个体分布图 Distribution of Individuals

径级分布表 DBH Class

胸径区间 (Diameter class) (cm)	个体数 (No. of individuals in the plot)	比例 (Proportion)
1～2	2689	0.25
2～5	5339	0.49
5～10	2615	0.24
10～20	145	0.01
20～30	1	0.00
30～60	4	0.00
≥60	0	0.00

124 映山红 yìngshānhóng | Sims Azalea

Rhododendron simsii Planch.

杜鹃花科 | Ericaceae

代码（SpCode）= RHOSIM

个体数（Individual number/24ha）= 4810

最大胸径（Max DBH）= 11.3 cm

重要值排序（Importance value rank）= 30

落叶灌木，高2～5m。分枝多而纤细，密被亮棕褐色扁平糙伏毛。叶常集生枝端，革质，两面被糙伏毛。花2～6朵簇生枝顶，玫瑰色，鲜红色或暗红色。蒴果密被糙伏毛。花期4～5月，果期6～8月。

Deciduous shrubs, 2-5 m tall; branches many and fine, densely shiny brown appressed-setose, setae flat. Leaves always clustered at apex of twiglets, blade leathery, coarsely appressed-hairy; Flowers 2-6 terminal clustered at twiglets, rose, bright to dark red. Capsule densely strigose. Fl. Apr.-May, fr. Jun.-Aug..

花枝 Flowering branch

摄影：金冬梅 Photo by: Jin Dongmei

花 Flower

摄影：陈彬 Photo by: Chen Bin

蒴果 Capsule

摄影：陈国科 Photo by: Chen Guoke

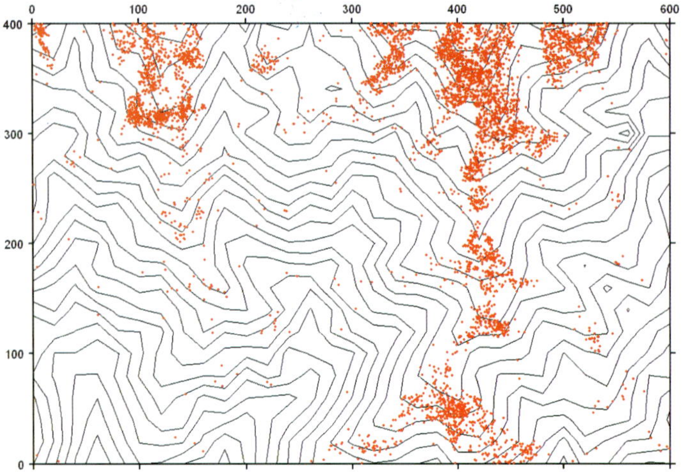

个体分布图 Distribution of Individuals

径级分布表 DBH Class

胸径区间 (Diameter class) (cm)	个体数 (No. of individuals in the plot)	比例 (Proportion)
1～2	3562	0.76
2～5	1153	0.24
5～10	2	0.00
10～20	2	0.00
20～30	1	0.00
30～60	0	0.00
≥60	0	0.00

125 南烛（乌饭、乌饭树） nánzhú(wūfàn、wūfànshù) | Oriental Blueberry

Vaccinium bracteatum Thunb.
杜鹃花科 | Ericaceae

代码（SpCode）= VACBRA
个体数（Individual number/24ha）= 2326
最大胸径（Max DBH）= 20 cm
重要值排序（Importance value rank）= 19

常绿灌木或小乔木，高2～9m。分枝多，幼枝被短柔毛或无毛，老枝紫褐色，无毛。叶片薄革质，侧脉5～7对。总状花序假顶生。浆果直径5～8mm，熟时紫黑色，有柔毛。花期6～7月，果期8～10月。

Evergreen shrub or small tree, 2-9 m tall, much branched. Twigs pubescent or glabrous; older branches purplish brown, glabrous. Leaf blade thinly leathery, lateral veins 5-7 pairs. Inflorescences pseudoterminal, racemose. Berry 5-8 mm, dark purple, pubescent. Fl. Jun.-Jul., fr. Aug.-Oct..

叶 Leaf
摄影：方腾 Photo by: Fang Teng

花 Flower
摄影：方腾 Photo by: Fang Teng

浆果 Berry
摄影：陈建华 Photo by: Chen Jianhua

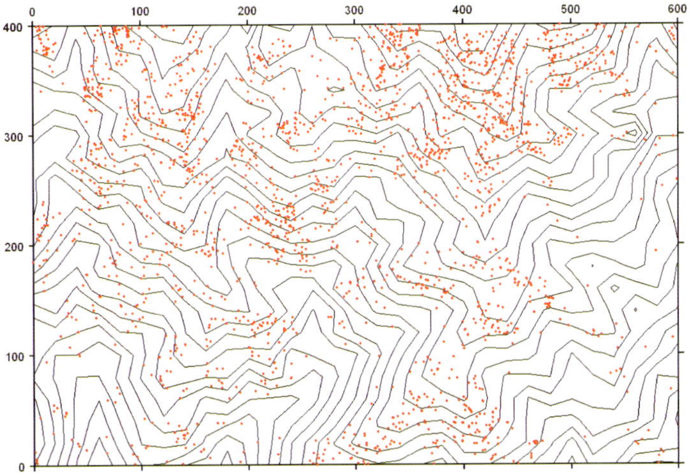

个体分布图 Distribution of Individuals

径级分布表 DBH Class

胸径区间 (Diameter class) (cm)	个体数 (No. of individuals in the plot)	比例 (Proportion)
1～2	1022	0.44
2～5	1100	0.47
5～10	175	0.08
10～20	29	0.01
20～30	0	0.00
30～60	0	0.00
≥60	0	0.00

126 短尾越橘（小叶乌饭） duǎnwěiyuèjú(xiǎoyèwūfàn) | Shortail Blueberry

Vaccinium carlesii Dunn

杜鹃花科 | Ericaceae

代码（SpCode）= VACCAR

个体数（Individual number/24ha）= 1803

最大胸径（Max DBH）= 15.7 cm

重要值排序（Importance value rank）= 14

常绿灌木或乔木，高1～6m。分枝多，纤细。幼枝被短柔毛，后变无毛。叶片革质，顶端渐尖。总状花序顶生和腋生。浆果球形，熟时紫黑色，无毛，常被白粉。花期5～6月，果期8～10月。

Evergreen shrub or small tree, 1-6 m tall, much branched, slender. Twigs slightly pubescent, glabrescent; Leaf blade leathery, apex acuminate. Inflorescences racemose, terminally or axillary. Berry globose, dark purple, glabrous, glaucous. Fl. May-Jun., fr. Aug.-Oct..

树干 Trunk

摄影：方腾 Photo by: Fang Teng

花序 Inflorescence

摄影：方腾 Photo by: Fang Teng

果枝 Fruiting branch

摄影：张金龙 Photo by: Zhang Jinlong

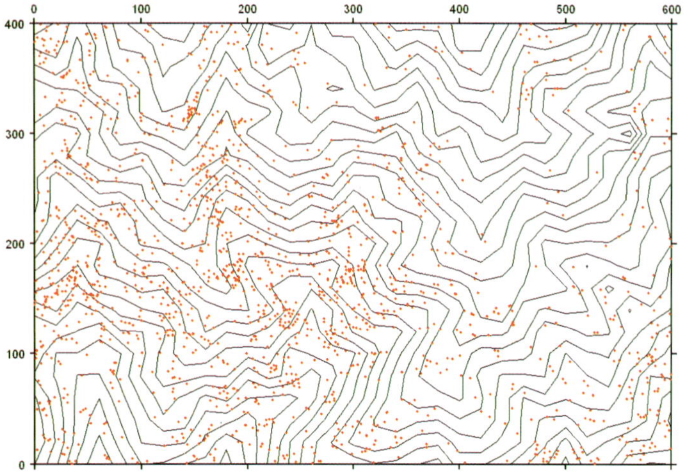

个体分布图 Distribution of Individuals

径级分布表 DBH Class

胸径区间 (Diameter class) (cm)	个体数 (No. of individuals in the plot)	比例 (Proportion)
1～2	508	0.28
2～5	945	0.52
5～10	338	0.19
10～20	12	0.01
20～30	0	0.00
30～60	0	0.00
≥60	0	0.00

127 江南越橘（米饭花） jiāngnányuèjú(mǐfànhuā) | China Blueberry

Vaccinium mandarinorum Diels
杜鹃花科 | Ericaceae

代码（SpCode）= VACMAN
个体数（Individual number/24ha）= 2232
最大胸径（Max DBH）= 14.2 cm
重要值排序（Importance value rank）= 11

常绿灌木或小乔木，高1～4m。幼枝通常无毛，有时被短柔毛，老枝紫褐色或灰褐色，无毛。叶片厚革质。总状花序腋生。浆果熟时紫黑色，无毛，直径4～6mm。花期4～6月，果期6～10月。

Evergreen shrub or small tree, 1-4 m tall. Young twigs glabrous or pubescent; older twigs purplish brown or greyish brown, glabrous. Leaf blade leathery. Inflorescences racemose, axillary. Berry dark purple, glabrous, 4-6 mm in diam. Fl. Apr.-Jun., fr. Jun.-Oct..

小枝 Branchlet
摄影：张金龙 Photo by: Zhang Jinlong

叶背 Leaf back
摄影：陈国科 Photo by: Chen Guoke

花序 Inflorescence
摄影：金冬梅 Photo by: Jin Dongmei

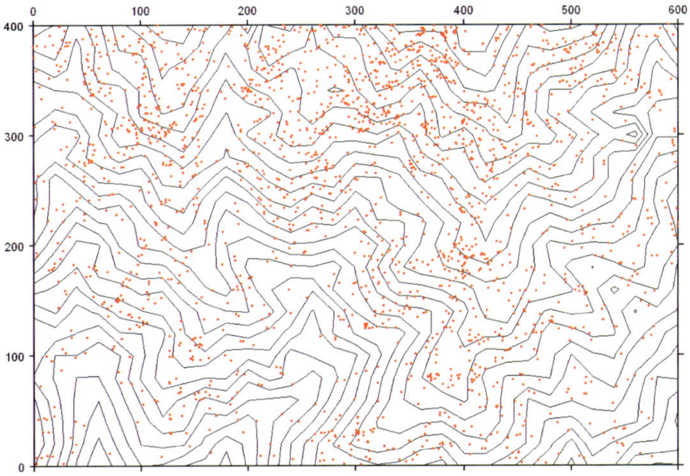

个体分布图 Distribution of Individuals

径级分布表 DBH Class

胸径区间 (Diameter class) (cm)	个体数 (No. of individuals in the plot)	比例 (Proportion)
1～2	1131	0.51
2～5	1061	0.48
5～10	37	0.02
10～20	3	0.00
20～30	0	0.00
30～60	0	0.00
≥60	0	0.00

128 朱砂根 zhūshāgēn | Hen's Eyes; Coral Berry; Hilo Holly

Ardisia crenata Sims
紫金牛科 | Myrsinaceae

代码（SpCode）= ARDCRE
个体数（Individual number/24ha）= 2
最大胸径（Max DBH）= 1.1 cm
重要值排序（Importance value rank）= 146

常绿灌木，高1～3m。茎粗壮，无毛，除侧生特殊花枝外，无分枝。叶片革质或纸质，边缘具明显腺点。伞形花序或聚伞花序生于侧生特殊花枝顶端。果实球形，鲜红色。花期5～6月，果期10月至翌年5月。

Evergreen shrub 1-3 m tall. Stem strong without lateral branchs other than specialized flowering twigs. Leaf blade leathery or papery, with glandular dots on margin. Inflorescences terminal, umbellate or cymose, on specialized, 2- or 3-leaved lateral branches. Fruit globose, red. Fl. May-Jun., fr. Oct. to May of next year.

果序 Infructescence
摄影：陈彬 Photo by: Chen Bin

叶背 Leaf back
摄影：张金龙 Photo by: Zhang Jinlong

植株 Whole plant
摄影：陈彬 Photo by: Chen Bin

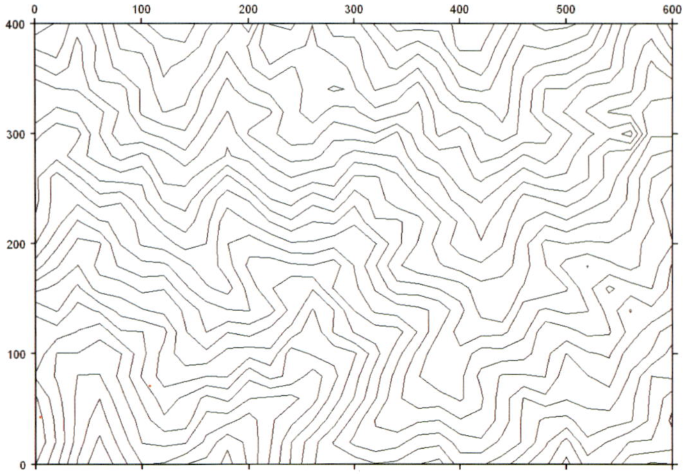

个体分布图 Distribution of Individuals

径级分布表 DBH Class

胸径区间 (Diameter class) (cm)	个体数 (No. of individuals in the plot)	比例 (Proportion)
1～2	2	1.00
2～5	0	0.00
5～10	0	0.00
10～20	0	0.00
20～30	0	0.00
30～60	0	0.00
≥60	0	0.00

129 山柿（粉叶柿、浙江柿） shānshì(zhèjiāngshì、fěnyèshì) | Zhejiang Persimmon

Diospyros japonica Sieb. et Zucc.
柿树科 | Ebenaceae

代码（SpCode）= DIOJAP
个体数（Individual number/24ha）= 140
最大胸径（Max DBH）= 31.1 cm
重要值排序（Importance value rank）= 59

落叶乔木，高达17m。树皮灰黑色或灰褐色，小枝散生纵裂的唇形小皮孔。叶革质，上面深绿色，下面粉绿色，侧脉7～9对。花雌雄异株。浆果黄红色，熟时红色，被白霜。花期4～7月，果期9～10月。

Deciduous tree, up to 17 m tall. Bark greyish black or greyish brown. Branchlets with scattered longitudinal split lenticels; leaf blade thinly leathery, lateral veins 7-9 pairs, adaxially dark green, abaxially pastal green. Flowers dioecious. Berries orange-yellow, becoming red, glaucous. Fl. Apr.-Jul., fr. Sep.-Oct..

小枝 Branchlet
摄影：宋凯 Photo by: Song Kai

花 Flower
摄影：方腾 Photo by: Fang Teng

浆果 Berry
摄影：方腾 Photo by: Fang Teng

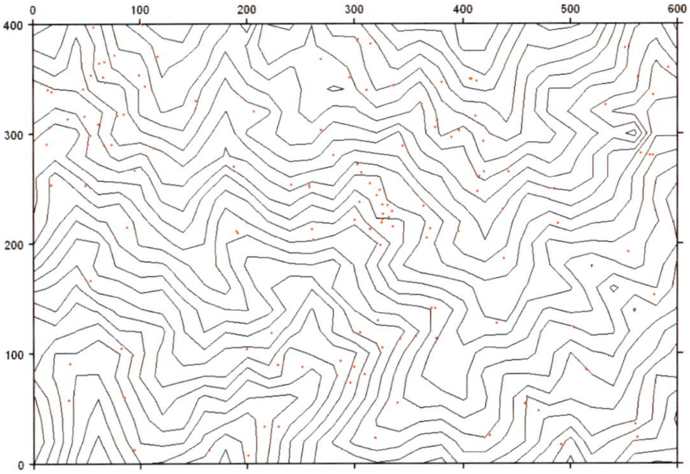

个体分布图 Distribution of Individuals

径级分布表 DBH Class

胸径区间 (Diameter class) (cm)	个体数 (No. of individuals in the plot)	比例 (Proportion)
1～2	12	0.09
2～5	13	0.09
5～10	16	0.11
10～20	46	0.33
20～30	50	0.36
30～60	3	0.02
≥60	0	0.00

130 罗浮柿 luófúshì | Morris Persimmon

Diospyros morrisiana Hance
柿树科 | Ebenaceae

代码（SpCode）= DIOMOR
个体数（Individual number/24ha）= 53
最大胸径（Max DBH）= 15.4 cm
重要值排序（Importance value rank）= 78

常绿灌木或乔木。高达20m。树皮呈片状剥落，表面黑色。叶薄革质，上面深绿色，下面绿色。侧脉4～8对。浆果黄色，球形。花期5～6月，果期11月。

Evergreen shrub or tree, up to 20 m tall. Bark peeling off in thin pieces, surface black. Leaf blade thinly leathery, adaxially dark green, abaxially green; lateral veins 4-8 pairs. Berries yellow, globose. Fl. May-Jun., fr. Nov..

幼苗 Seedling
摄影：金冬梅 Photo by: Jin Dongmei

叶 Leaf
摄影：陈国科 Photo by: Chen Guoke

果枝 Fruiting branch
摄影：陈国科 Photo by: Chen Guoke

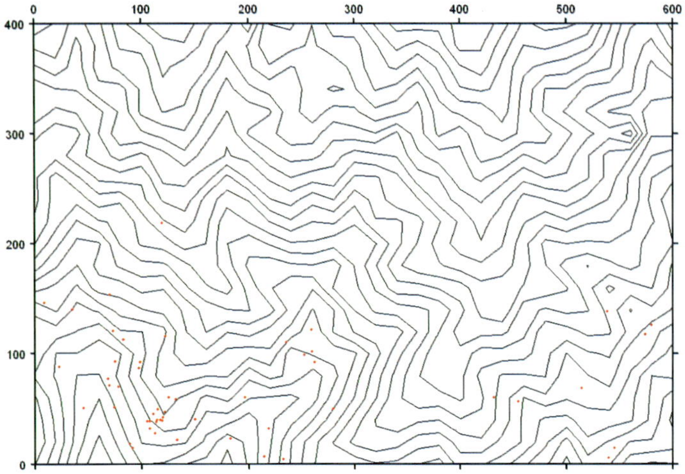

个体分布图 Distribution of Individuals

径级分布表 DBH Class

胸径区间 (Diameter class) (cm)	个体数 (No. of individuals in the plot)	比例 (Proportion)
1～2	12	0.23
2～5	19	0.36
5～10	13	0.25
10～20	9	0.17
20～30	0	0.00
30～60	0	0.00
≥60	0	0.00

131 薄叶山矾 báoyèshānfán | Thinleaf Sweetleaf

Symplocos anomala Brand
山矾科 | Symplocaceae

代码（SpCode）= SYMANO
个体数（Individual number/24ha）= 253
最大胸径（Max DBH）= 11 cm
重要值排序（Importance value rank）= 61

常绿灌木或小乔木，高达5m。顶芽、嫩枝被褐色柔毛。老枝通常黑褐色。叶薄革质，侧脉5～11对。总状花序野生，长8～15mm。核果褐色，长圆形，顶端有直立宿存萼片。花期4～12月，边开花边结果。

Evergreen shrub or small tree, up to 5 m tall. Apical bud and young branchlets tomentellous-tomentose to brown pubescent. Leaf blade thinly leathery, lateral veins 5-11 pairs. Racemes 0.8-1.5 cm. Drupes brown, oblong-globose, apex with persistent erect calyx lobes. Fl. and fr. Apr.-Dec..

幼苗 Seedling
摄影：金冬梅 Photo by: Jin Dongmei

叶 Leaf
摄影：张金龙 Photo by: Zhang Jinlong

花序 Inflorescence
摄影：方腾 Photo by: Fang Teng

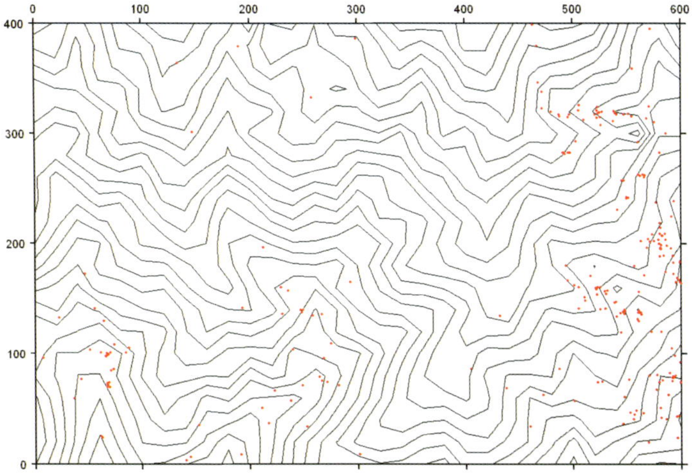

个体分布图 Distribution of Individuals

径级分布表 DBH Class

胸径区间 (Diameter class) (cm)	个体数 (No. of individuals in the plot)	比例 (Proportion)
1～2	96	0.38
2～5	128	0.51
5～10	28	0.11
10～20	1	0.00
20～30	0	0.00
30～60	0	0.00
≥60	0	0.00

132 山矾

shānfán | Sweetleaf

Symplocos sumuntia Buchanan-Hamilton ex D. Don

山矾科 | Symplocaceae

代码（SpCode）= SYMSUM

个体数（Individual number/24ha）= 56

最大胸径（Max DBH）= 7.6 cm

重要值排序（Importance value rank）= 87

常绿乔木，高达10m。嫩枝褐色，常光滑。叶薄革质，侧脉4～8对。总状花序长1～6cm。核果卵状坛形，长6～10mm，顶端宿存直立萼片。花期2～3月，果期6～7月。

Evergreen tree, up to 10 m tall. Young branchlets brown, usually glabrous. Leaf blade thinly leathery, lateral veins 4-8 pairs. Racemes 1-6 cm. Drupes ampulliform to ovoid, 6-10 mm, apex with persistent erect calyx lobes. Fl. Feb.-Mar., fr. Jun.-Jul..

幼苗 Seedling

摄影：金冬梅 Photo by: Jin Dongmei

花序 Inflorescence

摄影：杨波 Photo by: Yang Bo

果序 Infructescence

摄影：杨波 Photo by: Yang Bo

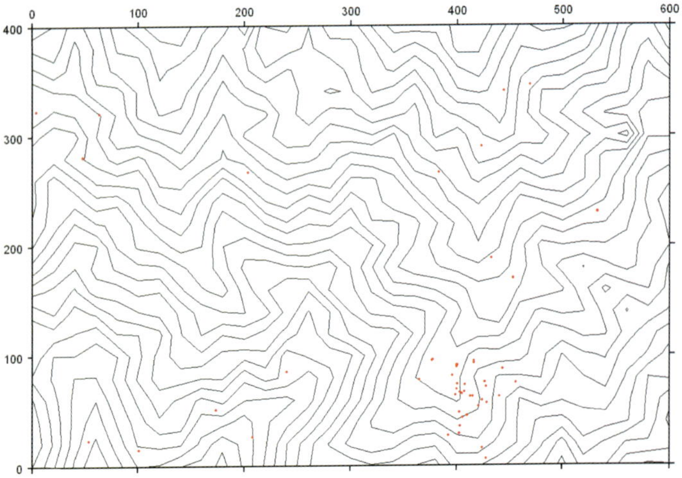

个体分布图 Distribution of Individuals

径级分布表 DBH Class

胸径区间 (Diameter class) (cm)	个体数 (No. of individuals in the plot)	比例 (Proportion)
1～2	21	0.38
2～5	31	0.55
5～10	4	0.07
10～20	0	0.00
20～30	0	0.00
30～60	0	0.00
≥60	0	0.00

133 白檀 báitán | Sapphireberry Sweetleaf

Symplocos paniculata (Thunb.) Miq.
山矾科 | Symplocaceae

代码（SpCode）= SYMPAN
个体数（Individual number/24ha）= 9
最大胸径（Max DBH）= 4.3 cm
重要值排序（Importance value rank）= 122

落叶灌木或小乔木，高达5m。嫩枝光滑或有灰白色柔毛，老枝无毛。叶膜质或薄纸质，侧脉4～8对。聚伞花序组成顶生圆锥花序，长2～10cm。核果熟时蓝色，卵状球形，稍偏斜。花期4～6月，果期9～11月。

Deciduous shrub or small tree, up to 5 m tall. Young branchlets glabrous or pilose; old branchlets glabrous. Leaf blade membranous to thinly papery, lateral veins 4-8 pairs. Panicles terminal, consisting of a raceme of cymose corymbs, 2-10 cm. Drupes bluish, rarely white, ovate-globose, little slanted. Fl. Apr.-Jun., fr. Sep.-Nov..

小枝 Branchlet
摄影：宋凯 Photo by: Song Kai

花枝 Flowering branch
摄影：陈彬 Photo by: Chen Bin

果实 Fruit
摄影：陈彬 Photo by: Chen Bin

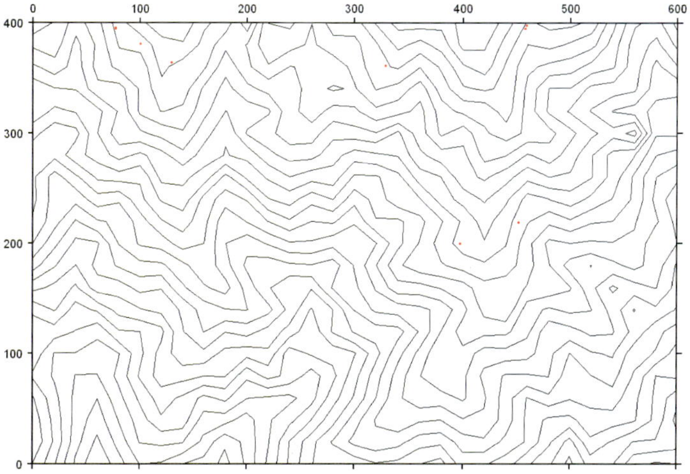

个体分布图 Distribution of Individuals

径级分布表 DBH Class

胸径区间 (Diameter class) (cm)	个体数 (No. of individuals in the plot)	比例 (Proportion)
1～2	6	0.67
2～5	3	0.33
5～10	0	0.00
10～20	0	0.00
20～30	0	0.00
30～60	0	0.00
≥60	0	0.00

134 光亮山矾（四川山矾） guāngliàngshānfán(sìchuānshānfán) | Sichuan Sweetleaf

Symplocos lucida (Thunberg) Sieb. et Zucc.
山矾科 | Symplocaceae

代码（SpCode）= SYMLUC
个体数（Individual number/24ha）= 17
最大胸径（Max DBH）= 13.4 cm
重要值排序（Importance value rank）= 115

常绿灌木或小乔木，高达5m。小枝黄绿色，有棱，无毛。叶革质，全缘或有尖锯齿。穗状花序缩短呈团伞状，或伸长达6cm。核果卵圆形或长圆形。花期3～11月，果期5～12月。

Evergreen shrub or small tree, up to 5 m tall. Branchlets mostly yellowish green, slightly ridged to winged, glabrous. Leaf blade leathery, margin entire or sharply serrate. Inflorescences a basally branched spike or raceme, shortened to a fascicle or up to 6 cm. Drupes ovoid to mostly ellipsoid. Fl. Mar.-Nov., fr. May.-Dec..

小枝 Branchlet
摄影：陈彬 Photo by: Chen Bin

幼苗 Seedling
摄影：金冬梅 Photo by: Jin Dongme

核果 Drupe
摄影：陈彬 Photo by: Chen Bin

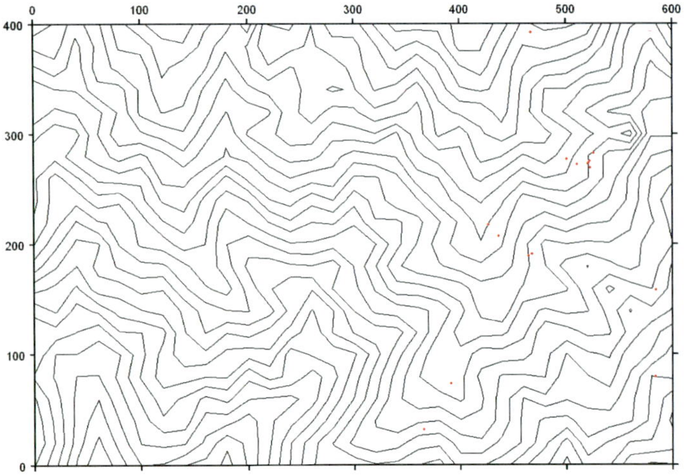

个体分布图 Distribution of Individuals

径级分布表 DBH Class

胸径区间 (Diameter class) (cm)	个体数 (No. of individuals in the plot)	比例 (Proportion)
1～2	3	0.18
2～5	10	0.59
5～10	2	0.12
10～20	2	0.12
20～30	0	0.00
30～60	0	0.00
≥60	0	0.00

135 老鼠矢 láoshǔshǐ | Starshape Sweetleaf

Symplocos stellaris Brand
山矾科 | Symplocaceae

代码（SpCode）= SYMSTE
个体数（Individual number/24ha）= 629
最大胸径（Max DBH）= 13.4 cm
重要值排序（Importance value rank）= 28

常绿灌木或小乔木，高达8m。小枝粗，髓心中空。芽、嫩枝、嫩叶柄均被红褐色绒毛。叶片狭长，厚革质。团伞花序生于二年生枝的叶痕上。核果狭卵状圆柱形。花期2～5月，果期6～8月。

Evergreen shrub or small tree, up to 8 m tall. Twiglets stout, pith empty. Buds, petioles and young branchlets reddish brown tomentellous. Leaf blade narrow, thickly leathery. Inflorescences a glomerule or condensed spike, at leaf scars of second year branchlets. Drupes narrowly ovoid-cylindric. Fl. Feb.-May, fr. Jun.-Aug..

嫩枝 New branch
摄影：方腾 Photo by: Fang Teng

叶 Leaf
摄影：张金龙 Photo by: Zhang Jinlong

花 Flower
摄影：方腾 Photo by: Fang Teng

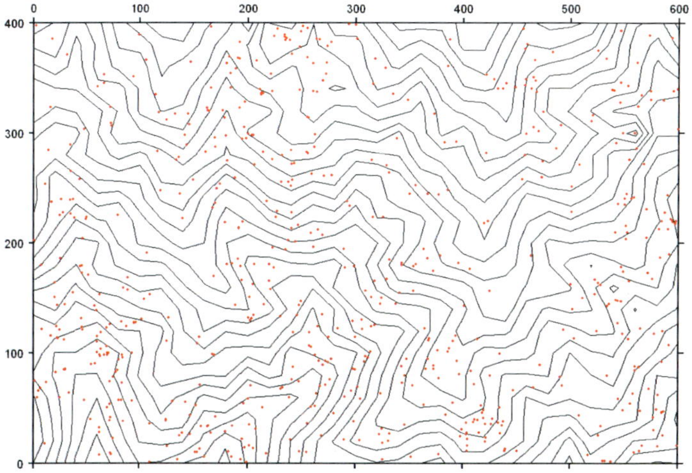

个体分布图 Distribution of Individuals

径级分布表 DBH Class

胸径区间 (Diameter class) (cm)	个体数 (No. of individuals in the plot)	比例 (Proportion)
1～2	243	0.39
2～5	289	0.46
5～10	87	0.14
10～20	10	0.02
20～30	0	0.00
30～60	0	0.00
≥60	0	0.00

136 赤杨叶

chìyángyè | Fortune Chinabells

Alniphyllum fortunei (Hemsl.) Makino

安息香科 | Styracaceae

代码（SpCode）= ALNFOR

个体数（Individual number/24ha）= 247

最大胸径（Max DBH）= 38.5 cm

重要值排序（Importance value rank）= 47

落叶乔木，高达20m。树干通直，树皮不开裂。叶膜质到纸质。总状花序或圆锥花序，顶生或腋生。果实长圆形或长椭圆形，成熟时5瓣开裂。种子两端有膜质翅。花期4～7月，果期8～10月。

Deciduous trees, up to 20 m tall. Trunk straight; bark not cracked. Leaf blade membranous to papery. Inflorescence raceme or panicle, terminal or axillary. Fruit oblong to ellipsoid, dehiscence to 5 lobes. Seeds with membranous wings at both ends. Fl. Apr.-Jul., fr. Aug.-Oct..

幼苗 Seedling

摄影：金冬梅 Photo by: Jin Dongmei

花序 Inflorescence

摄影：方腾 Photo by: Fang Teng

果枝 Fruiting branch

摄影：陈国科 Photo by: Chen Guoke

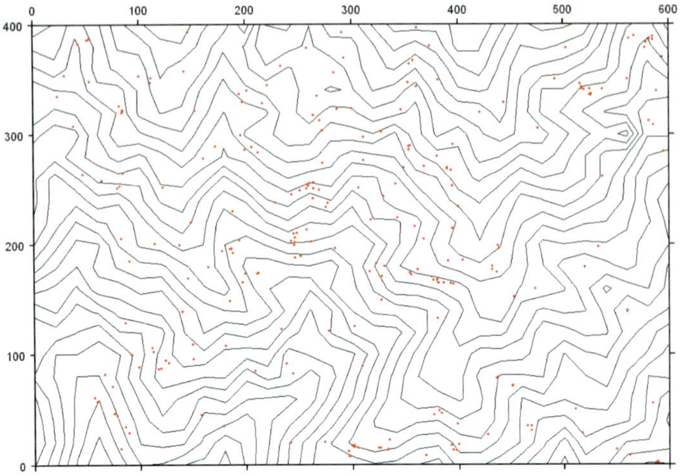

个体分布图 Distribution of Individuals

径级分布表 DBH Class

胸径区间 (Diameter class) (cm)	个体数 (No. of individuals in the plot)	比例 (Proportion)
1～2	49	0.20
2～5	64	0.26
5～10	48	0.19
10～20	53	0.21
20～30	27	0.11
30～60	6	0.02
≥60	0	0.00

137 垂珠花 chuízhūhuā | Roughhairflower Storax

Styrax dasyanthus Perkins
安息香科 | Styracaceae

代码（SpCode）= STYDAS
个体数（Individual number/24ha）= 113
最大胸径（Max DBH）= 12.9 cm
重要值排序（Importance value rank）= 65

落叶乔木，高3～20m。嫩枝密被灰黄色星状微柔毛，成长后无毛，紫红色。叶革质或近革质。圆锥花序或总状花序顶生或腋生。果实密被黄色短绒毛。花期3～5月，果期9～12月。

Deciduous tree 3-20 m tall. Branchlets densely gray-yellow stellate pubescent, becoming purple and glabrous. Leaf blade leathery to almost leathery. Inflorescences terminal or axillary, paniculate. Fruit densely grayish stellate tomentose. Fl. Mar.-May, fr. Sep.-Dec..

蒴果 Capsule
摄影：陈国科 Photo by: Chen Guoke

幼苗 Seedling
摄影：金冬梅 Photo by: Jin Dongmei

花枝 Flowering branch
摄影：方腾 Photo by: Fang Teng

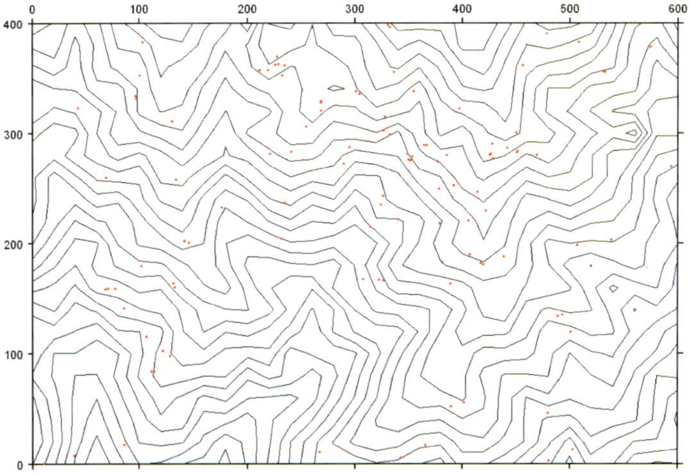

个体分布图 Distribution of Individuals

径级分布表 DBH Class

胸径区间 (Diameter class) (cm)	个体数 (No. of individuals in the plot)	比例 (Proportion)
1～2	43	0.38
2～5	34	0.30
5～10	28	0.25
10～20	8	0.07
20～30	0	0.00
30～60	0	0.00
≥60	0	0.00

138 栓叶安息香（红皮树） shuānyèānxīxiāng(hóngpíshù) | Corkleaf Snowbell

Styrax suberifolius Hook. et Arn.
安息香科 | Styracaceae

代码（SpCode）= STYSUB
个体数（Individual number/24ha）= 3
最大胸径（Max DBH）= 12.6 cm
重要值排序（Importance value rank）= 136

常绿乔木，高达20m。树皮红褐色或灰褐色。嫩枝被锈褐色星状绒毛。老枝无毛，紫褐色或灰褐色。叶革质，下面密被星状绒毛。花序顶生或腋生。果实卵状球形，熟时三裂。花期3～5月，果期9～11月。

Evergreen tree, up to 20 m tall. Bark redish brown or greyish brown. Branchlets red-brown to gray-brown stellate tomentose. Older branches glabrous, purplish brown or grayish brown. Leaf blade leathery, abaxially densely brownish stellate tomentose. Inflorescences terminal or axillary. Fruit ovoid-globose, 3-valvate dehiscent. Fl. Mar.-May, fr. Sep.-Nov..

树干 Trunk
摄影：贾琪 Photo by: Jia Qi

叶面 Leaf upper surface
摄影：贾琪 Photo by: Jia Qi

叶背 Leaf back
摄影：贾琪 Photo by: Jia Qi

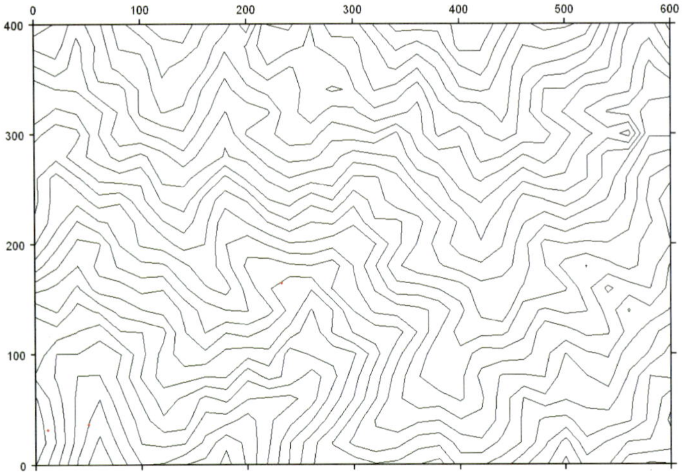

个体分布图 Distribution of Individuals

径级分布表 DBH Class

胸径区间 (Diameter class) (cm)	个体数 (No. of individuals in the plot)	比例 (Proportion)
1～2	1	0.33
2～5	0	0.00
5～10	1	0.33
10～20	1	0.33
20～30	0	0.00
30～60	0	0.00
≥60	0	0.00

139 芬芳安息香（郁香野茉莉） fēnfāngānxīxiāng(yùxiāngyěmòlì) | Sweetscented Snowbell

Styrax odoratissimus Champ.
安息香科 | Styracaceae

代码（SpCode）= STYODO
个体数（Individual number/24ha）= 537
最大胸径（Max DBH）= 16.7 cm
重要值排序（Importance value rank）= 37

落叶小乔木，高4～10m。树皮灰褐色，不开裂。叶革质至纸质。总状或圆锥花序，顶生。果实近球形，密被灰黄色星状绒毛。种子卵形，密被鳞片状毛和瘤状突起。花期3～4月，果期6～9月。

Deciduous small tree, 4-10 m tall. Bark gray-brown, not exfoliating. Leaf blade leathery to papery. Inflorescence terminal racemes or panicles. Fruit subglobose, densely gray-yellow stellate tomentose. Seeds ovoid, tuberculate, densely brown scaly. Fl. Mar.-Apr., fr. Jun.-Sep..

果序 Infructescence
摄影：方腾 Photo by: Fang Teng

叶 Leaf
摄影：方腾 Photo by: Fang Teng

花枝 Flowering branch
摄影：陈彬 Photo by: Chen Bin

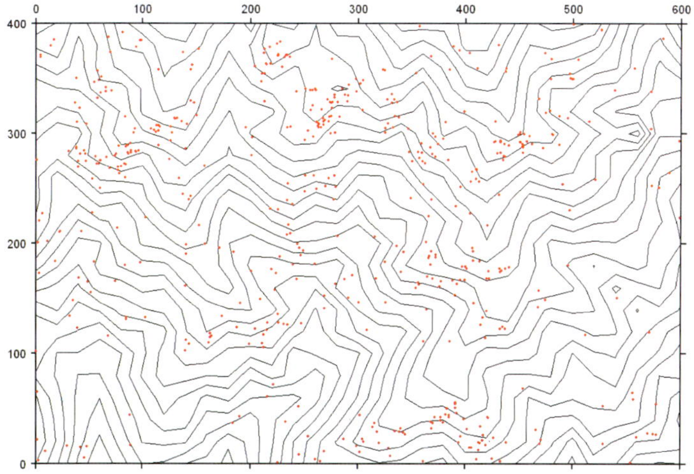

个体分布图 Distribution of Individuals

径级分布表 DBH Class

胸径区间 (Diameter class) (cm)	个体数 (No. of individuals in the plot)	比例 (Proportion)
1～2	206	0.38
2～5	208	0.39
5～10	105	0.20
10～20	18	0.03
20～30	0	0.00
30～60	0	0.00
≥60	0	0.00

140 苦枥木

kǔlìmù | Insular Ash

Fraxinus insularis Hemsl.

木犀科 | Oleaceae

代码（SpCode）= FRAINS

个体数（Individual number/24ha）= 513

最大胸径（Max DBH）= 47.1 cm

重要值排序（Importance value rank）= 33

落叶乔木，高3～20m。芽狭三角状圆锥形，密被黑褐色绒毛，干后黑色。羽状复叶对生，叶轴基部膨大，小叶3～7枚，硬纸质。圆锥花序顶生或腋生枝梢。翅果匙形。花期4～5月，果期7～9月。

Deciduous tree, 3-20 m tall. Buds narrowly conical, densely dark-brown tomentose, becoming shiny black when dry. Compound leaves opposite, axis base inflated; leaflets 3-7, blade papery to somewhat leathery. Panicles terminal or lateral. Samara spatulate. Fl. Apr.-May., fr. Jul.-Sep..

幼苗 Seedling

摄影：方腾 Photo by: Fang Teng

叶 Leaf

摄影：陈建华 Photo by: Chen Jianhua

花序 Inflorescence

摄影：方腾 Photo by: Fang Teng

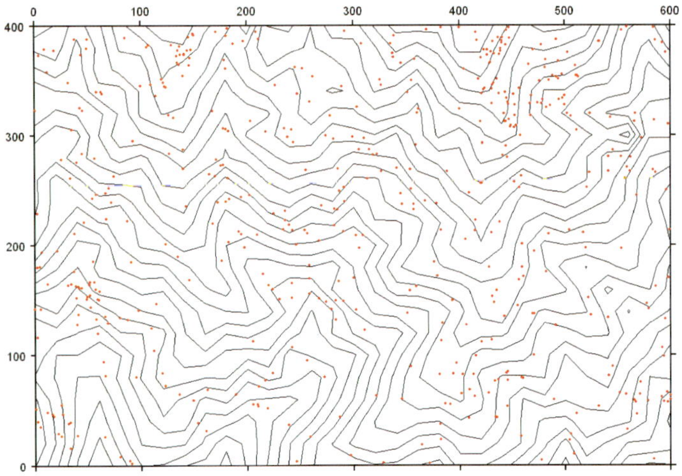

个体分布图 Distribution of Individuals

径级分布表 DBH Class

胸径区间 (Diameter class) (cm)	个体数 (No. of individuals in the plot)	比例 (Proportion)
1～2	204	0.40
2～5	89	0.17
5～10	43	0.08
10～20	115	0.22
20～30	52	0.10
30～60	10	0.02
≥60	0	0.00

141 宁波木犀（华东木犀） níngbōmùxī(huádōngmùxī) | Cooper Osmanther

Osmanthus cooperi Hemsl.
木犀科 | Oleaceae

代码（SpCode）= OSMCOO
个体数（Individual number/24ha）= 185
最大胸径（Max DBH）= 21.3 cm
重要值排序（Importance value rank）= 60

常绿小乔木或灌木，高3～8m。小枝灰白色。叶片革质，全缘，侧脉7～8对，在两面均极不明显。花序簇生于叶腋，每腋内有花4～12朵。核果长圆形，熟时蓝黑色。花期9～10月，果期翌年5～6月。

Evergreen shrub or small tree 3-8 m tall. Twiglets greyish white. Leaf blade leathery, margin entire, lateral veins 7-8 pairs, obscure. Cymes fascicled in leaf axils, 4-12-flowered. Drupe oblong, blue-black at maturity. Fl. Sep.-Oct., fr. May-Jun. of next year.

幼苗 Seedling
摄影：邴艳红 Photo by: Bing Yanhong

叶背 Leaf back
摄影：陈国科 Photo by: Chen Guoke

花序 Inflorescence
摄影：方腾 Photo by: Fang Teng

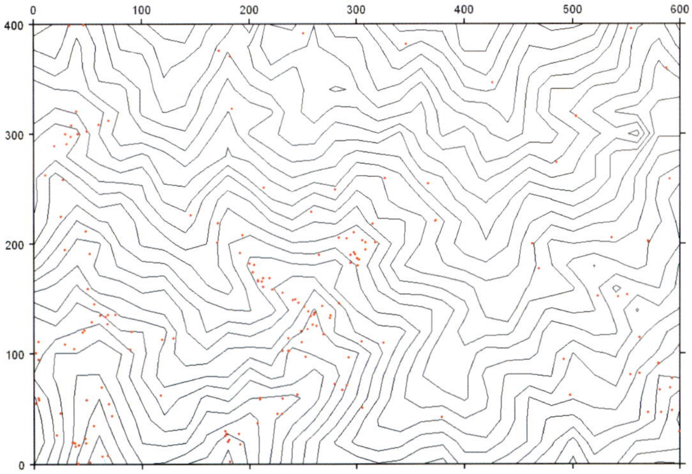

个体分布图 Distribution of Individuals

径级分布表 DBH Class

胸径区间 (Diameter class) (cm)	个体数 (No. of individuals in the plot)	比例 (Proportion)
1～2	26	0.14
2～5	49	0.26
5～10	80	0.43
10～20	29	0.16
20～30	1	0.01
30～60	0	0.00
≥60	0	0.00

142 厚壳树 hòukéshù | Koda tree

Ehretia acuminata (DC.) R. Br.
紫草科 | Boraginaceae

代码（SpCode）= EHRACU
个体数（Individual number/24ha）= 1
最大胸径（Max DBH）= 1.2 cm
重要值排序（Importance value rank）= 157

落叶乔木，高达15m。树皮灰黑色，纵裂。小枝褐色，光滑，有明显皮孔。叶纸质，边缘有整齐的内弯锯齿。聚伞花序圆锥状，芳香。核果黄色或橘黄色。花期4～5月，果熟期7～8月。

Deciduous tree, up to 15 m tall; bark grayish black, laciniate; branchlets brown, glabrous, with distinct lenticels; Leaf blade papery, margin regularly serrate with teeth curved upward. Flowers crowded, fragrant. Drupes yellow or orange. Fl. Apr.-May, fr. Jul.-Aug..

幼苗 Seedling
摄影：邴艳红 Photo by: Bing Yanhong

叶 Leaf
摄影：陈建华 Photo by: Chen Jianhua

花序 Inflorescence
摄影：方腾 Photo by: Fang Teng

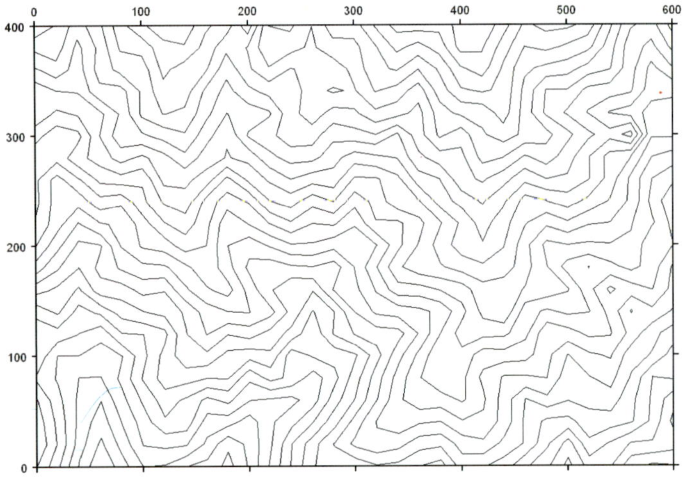

个体分布图 Distribution of Individuals

径级分布表 DBH Class

胸径区间 (Diameter class) (cm)	个体数 (No. of individuals in the plot)	比例 (Proportion)
1～2	1	1.00
2～5	0	0.00
5～10	0	0.00
10～20	0	0.00
20～30	0	0.00
30～60	0	0.00
≥60	0	0.00

143 紫珠 zǐzhū | Purple Pearl

Callicarpa bodinieri Lévl.
马鞭草科 | Verbenaceae

代码（SpCode）= CALBOD
个体数（Individual number/24ha）= 3
最大胸径（Max DBH）= 1.6 cm
重要值排序（Importance value rank）= 138

落叶灌木，高约2m。小枝、叶柄和花序被粗糠状星状毛。叶片具红色腺点，叶背近无毛或被星状柔毛，上面光滑。聚伞花序腋生。果实球形，熟时紫色，无毛。花期6～7月，果期8～11月。

Deciduous shrub about 2 m tall; branchlets, petioles, and inflorescences with dense stellate hairs. Leaf blade red glandular, abaxially subglabrous or stellate tomentose, adaxially pubescent. Cymes axillary. Fruit globose, purple, ca 2 mm in diam. Fl. Jun.-Jul., fr. Aug.-Nov..

花序 Inflorescence
摄影：邴艳红 Photo by: Bing Yanhong

幼苗 Seedling
摄影：金冬梅 Photo by: Jin Dongmei

果序 Infructescence
摄影：易思荣 Photo by: Yi Sirong

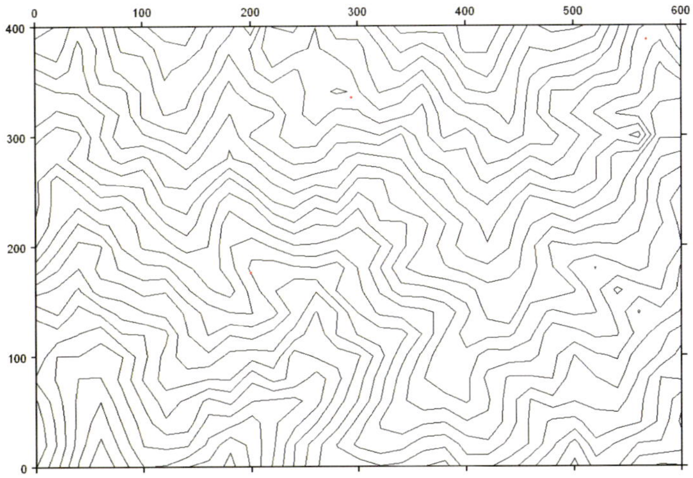

个体分布图 Distribution of Individuals

径级分布表 DBH Class

胸径区间 (Diameter class) (cm)	个体数 (No. of individuals in the plot)	比例 (Proportion)
1～2	3	1.00
2～5	0	0.00
5～10	0	0.00
10～20	0	0.00
20～30	0	0.00
30～60	0	0.00
≥60	0	0.00

144 老鸦糊 lǎoyāhú | Ly Purplepearl

Callicarpa giraldii Hesse ex Rehd.
马鞭草科 | Verbenaceae

代码（SpCode）= CALGIR
个体数（Individual number/24ha）= 2
最大胸径（Max DBH）= 1.2 cm
重要值排序（Importance value rank）= 145

落叶灌木，高1～5m。小枝灰黄色，被星状毛。叶片纸质，基部楔形或下延成楔形。聚伞花序腋生，被毛。果实球形，初时疏被星状毛，熟时无毛，紫色，直径2.5～4mm。花期5～7月，果期7～11月。

Deciduous shrub 1-5 m tall. Twiglets greyish yellow, with stellate hairs. Leaf blade papery, base cuneate, narrowly cuneate. Cymes axillary, stellate tomentose. Fruit purple, 2.5-4 mm in diam, stellate tomentose when young, glabrescent when mature, purple. Fl. May-Jul., fr. Jul.-Nov..

小枝 Branchlet
摄影：宋凯 Photo by: Song Kai

叶背 Leaf back
摄影：陈国科 Photo by: Chen Guoke

花枝 Flowering Brabchlet
摄影：张芬耀 Photo by: Zhang Fenyao

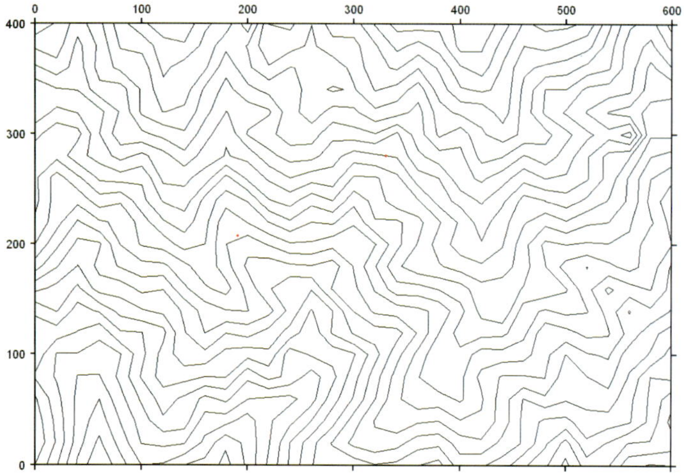

个体分布图 Distribution of Individuals

径级分布表 DBH Class

胸径区间 (Diameter class) (cm)	个体数 (No. of individuals in the plot)	比例 (Proportion)
1～2	2	1.00
2～5	0	0.00
5～10	0	0.00
10～20	0	0.00
20～30	0	0.00
30～60	0	0.00
≥60	0	0.00

145 红紫珠　hóngzǐzhū | Reddish Purplepearl

Callicarpa rubella Lindl.
马鞭草科 | Verbenaceae

代码（SpCode）= CALRUB
个体数（Individual number/24ha）= 1
最大胸径（Max DBH）= 1 cm
重要值排序（Importance value rank）= 159

落叶灌木，高约2m。小枝被黄褐色星状毛并杂有多细胞的腺毛。叶片纸质，基部心形，有时偏斜，叶柄极短或无柄。聚伞花序宽2～4cm。果实紫红色，直径约2mm。花期5～7月，果期7～11月。

Deciduous shrub about 2 m tall. Twiglets with greyish brown stellate hairs and multicell glandular hairs; leaf blade papery, base cordate and sometimes oblique; petiole very short or absent. Cymes 2-4 cm across. Fruit purple, ca. 2 mm in diam. Fl. May-Jul., fr. Jul.-Nov..

叶　Leaf
摄影：张金龙　Photo by: Zhang Jinlong

幼苗　Seedling
摄影：金冬梅　Photo by: Jin Dongmei

果序　Infructescence
摄影：方腾　Photo by: Fang Teng

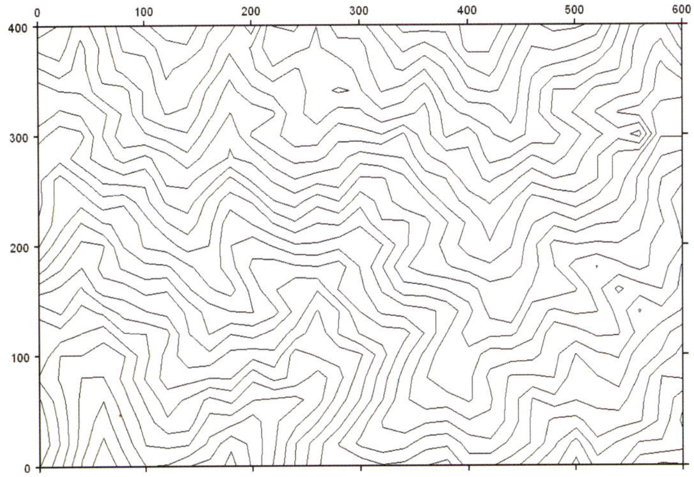

个体分布图 Distribution of Individuals

径级分布表 DBH Class

胸径区间 (Diameter class) (cm)	个体数 (No. of individuals in the plot)	比例 (Proportion)
1～2	1	1.00
2～5	0	0.00
5～10	0	0.00
10～20	0	0.00
20～30	0	0.00
30～60	0	0.00
≥60	0	0.00

146 大青

dàqīng | Manyflower Glorybower

Clerodendrum cyrtophyllum Turcz.
马鞭草科 | Verbenaceae

代码（SpCode）= CLECYR
个体数（Individual number/24ha）= 28
最大胸径（Max DBH）= 5.5 cm
重要值排序（Importance value rank）= 100

常绿灌木或小乔木，高1～10m。幼枝被短柔毛。叶片纸质，背面常有腺点。伞房状聚伞花序，生于枝顶或叶腋。结果时花萼宿存，红色。核果球形或倒卵形，绿色，成熟时蓝黑色。花果期6月至翌年2月。

Evergreen shrub or small tree, 1-10 m tall. Branchlets pubescent. Leaf blade papery, abaxially glandular. Inflorescences corymbose, formed from terminal and subterminal cymes. Fruiting calyx red. Drupes obovate to globose, green when immature, blue-purple at maturity. Fl. and fr. Jun.-Feb. of next year.

幼苗 Seedling
摄影：方腾 Photo by: Fang Teng

叶 Leaf
摄影：陈建华 Photo by: Chen Jianhua

果枝 Fruiting branch
摄影：陈彬 Photo by: Chen Bin

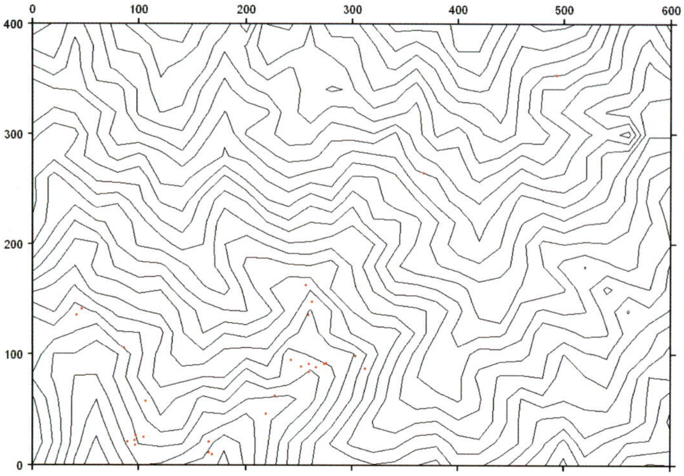

个体分布图 Distribution of Individuals

径级分布表 DBH Class

胸径区间 (Diameter class) (cm)	个体数 (No. of individuals in the plot)	比例 (Proportion)
1～2	9	0.32
2～5	17	0.61
5～10	2	0.07
10～20	0	0.00
20～30	0	0.00
30～60	0	0.00
≥60	0	0.00

147 浙江大青

zhèjiāngdàqīng | Zhejiang Glorybower

Clerodendrum kaichianum Hsu
马鞭草科 | Verbenaceae

代码（SpCode）= CLEKAI
个体数（Individual number/24ha）= 3
最大胸径（Max DBH）= 4 cm
重要值排序（Importance value rank）= 150

落叶灌木或小乔木，高2～5m。嫩枝近四棱形，密生黄褐色、褐色或红褐色短柔毛；老枝褐色。叶片厚纸质，全缘，两侧稍不对称，侧脉5～6对。伞房状聚伞花序顶生。核果蓝绿色。花果期7～11月。

Deciduous shrub or small tree, 2-5 m tall. Branchlets nearly 4-angled, densely yellow-brown, brown, or reddish brown pubescent when young. Leaf blade thickly papery, base slightly oblique, margin entire, veins 5 or 6 pairs. Inflorescences apical, corymbose cymes. Drupes blue-green. Fl. and fr. Jul.-Nov..

花序 Inflorescence
摄影：方腾 Photo by: Fang Teng

幼苗 Seedling
摄影：金冬梅 Photo by: Jin Dongmei

花 Flower
摄影：方腾 Photo by: Fang Teng

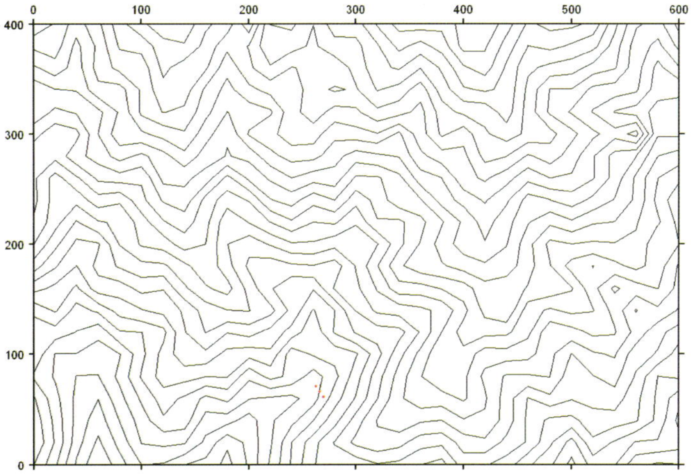

个体分布图 Distribution of Individuals

径级分布表 DBH Class

胸径区间 (Diameter class) (cm)	个体数 (No. of individuals in the plot)	比例 (Proportion)
1～2	0	0.00
2～5	3	1.00
5～10	0	0.00
10～20	0	0.00
20～30	0	0.00
30～60	0	0.00
≥60	0	0.00

148 豆腐柴（腐婢） dòufuchái(fŭbì) | Japan Premna

Premna microphylla Turcz.
马鞭草科 | Verbenaceae

代码（SpCode）= PREMIC
个体数（Individual number/24ha）= 21
最大胸径（Max DBH）= 2 cm
重要值排序（Importance value rank）= 98

直立灌木。幼枝有柔毛，老枝变无毛。叶揉之有臭味。叶片膜质，无柄或基部渐狭窄下延至叶柄两侧。聚伞花序组成顶生塔形的圆锥花序。核果紫色，球形至倒卵形。花果期5～10月。

Evergreen shrub, erect. Branchlets pubescent when young, glabrescent. Leaves odorous when crushed; leaf blade membranous, sessile or short petiolate with a winged petiole. Inflorescences conical panicles, terminal. Fruit purple, globose to obovate. Fl. and fr. May-Oct..

幼苗 Seedling
摄影：金冬梅 Photo by: Jin Dongmei

小枝 Branchlet
摄影：陈彬 Photo by: Chen Bin

花 Flower
摄影：陈彬 Photo by: Chen Bin

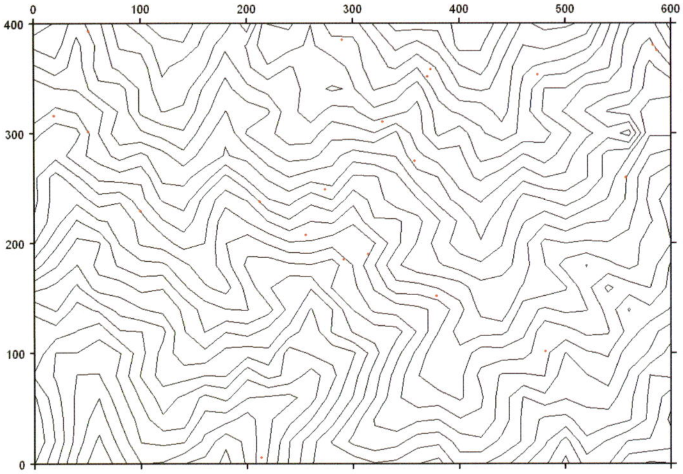

个体分布图 Distribution of Individuals

径级分布表 DBH Class

胸径区间 (Diameter class) (cm)	个体数 (No. of individuals in the plot)	比例 (Proportion)
1～2	21	1.00
2～5	0	0.00
5～10	0	0.00
10～20	0	0.00
20～30	0	0.00
30～60	0	0.00
≥60	0	0.00

149 栀子

zhīzi | Gardenia; Cape Jasmine

Gardenia jasminoides J. Ellis
茜草科 | Rubiaceae

代码（SpCode）= GARJAS
个体数（Individual number/24ha）= 204
最大胸径（Max DBH）= 3.2 cm
重要值排序（Importance value rank）= 55

常绿灌木，高0.3～3m。嫩枝常被短毛。叶对生，少为3枚轮生，革质。花芳香，通常单朵生于枝顶。花冠白色或乳黄色。果实黄色或橙红色，有翅状纵棱与宿存萼片。花期3～7月，果期5月到翌年2月。

Evergreen shrub, 0.3-3 m tall. Twiglets always pubescent. Leaves opposite or ternate, blade leathery. Flowers fragrant, corona white or butter yellow. Fruits yellow or orange, with longitudinal wings and permanent calyces. Fl. Mar.-Jul., fr. May to Feb. of next year.

幼苗 Seedling
摄影：金冬梅 Photo by: Jin Dongmei

花 Flower
摄影：杨波 Photo by: Yang Bo

果 Fruit
摄影：杨波 Photo by: Yang Bo

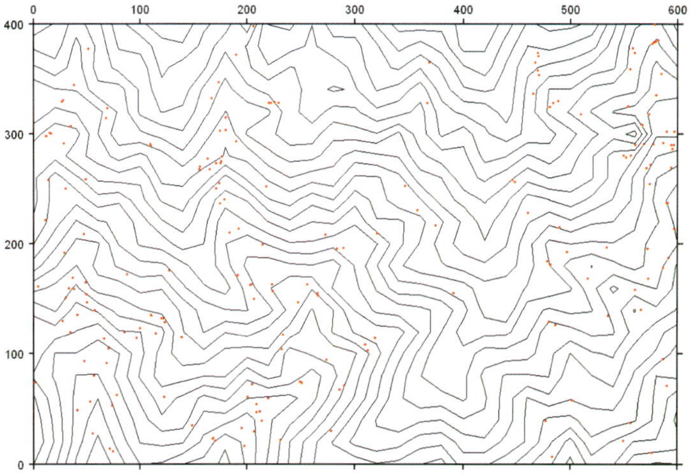

个体分布图 Distribution of Individuals

径级分布表 DBH Class

胸径区间 (Diameter class) (cm)	个体数 (No. of individuals in the plot)	比例 (Proportion)
1～2	184	0.90
2～5	20	0.10
5～10	0	0.00
10～20	0	0.00
20～30	0	0.00
30～60	0	0.00
≥60	0	0.00

150 日本粗叶木 rìběncūyèmù | Japan Roughleaf

Lasianthus japonicus Miq.
茜草科 | Rubiaceae

代码（SpCode）= LASJAP
个体数（Individual number/24ha）= 14
最大胸径（Max DBH）= 1.5 cm
重要值排序（Importance value rank）= 116

常绿灌木。枝和小枝无毛或嫩部被柔毛。叶近革质或纸质。花无梗，常2～3朵生在一个腋生、很短的总梗上，有时无总梗。核果球形，径约5mm，内含5个分核。花期5～6月，果期10～11月。

Evergreen shrub. Branchlets and twiglets glabrous or pubescent at young parts. Leaf blade almost leathery or papery. Flowers without pedicel, 2-3 clustered on axillary short stalk, sometimes without stalk. Drupe globose, ca. 5 mm in diam, with 5 pyrenes. Fl. May-Jun., fr. Oct.-Nov..

叶 Leaf
摄影：张金龙 Photo by: Zhang Jinlong

花 Flower
摄影：金冬梅 Photo by: Jin Dongmei

核果 Drupe
摄影：陈国科 Photo by: Chen Guoke

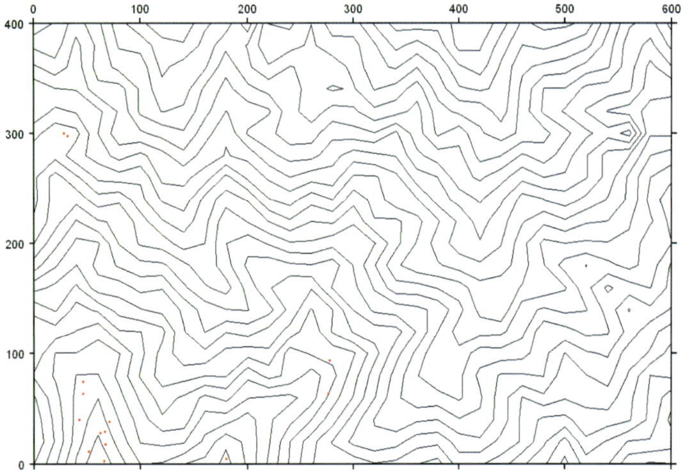

个体分布图 Distribution of Individuals

径级分布表 DBH Class

胸径区间 (Diameter class) (cm)	个体数 (No. of individuals in the plot)	比例 (Proportion)
1～2	14	1.00
2～5	0	0.00
5～10	0	0.00
10～20	0	0.00
20～30	0	0.00
30～60	0	0.00
≥60	0	0.00

151 海南槽裂木 hǎináncáolièmù | Pertusadina

Pertusadina hainanensis (F. C. How) Ridsdale
茜草科 | Rubiaceae

代码（SpCode）= PERHAI
个体数（Individual number/24ha）= 75
最大胸径（Max DBH）= 15.4 cm
重要值排序（Importance value rank）= 86

常绿灌木或乔木，高达30m。幼枝栗色，无毛或近无毛。叶厚纸质，侧脉7～10对。头状花序，花序梗单一，不分枝或偶尔二歧状分枝。花冠黄色，芳香。果序直径4～6mm。花期5～6月，果期7～10月。

Evergreen shrub or tree, up to 30 m tall. Twiglets maroon, glabrous or nearly so. Leaf blade thickly papery, lateral viens 7-10 pairs. Inflorescence capitate axillary, pedicel solitary or seldomly dichotomous. Flower corona yellow, fragrant. Infructescence 4-6 mm in diam. Fl. May-Jun., fr. Jul.-Oct..

幼苗 Seedling
摄影：方腾 Photo by: Fang Teng

叶面 Leaf upper surface
摄影：张金龙 Photo by: Zhang Jinlong

花序 Inflorescence
摄影：方腾 Photo by: Fang Teng

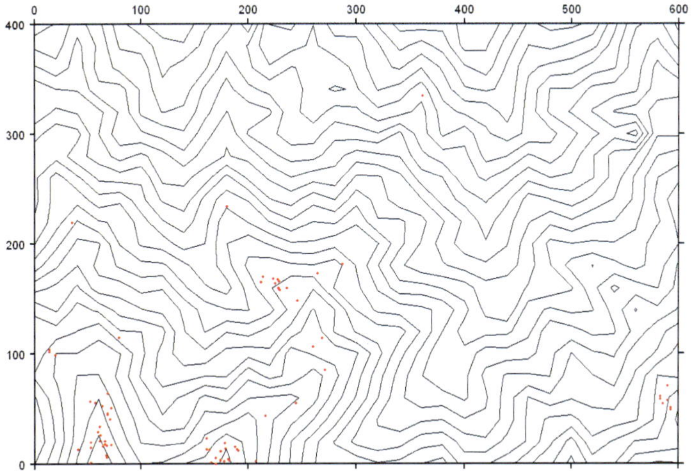

个体分布图 Distribution of Individuals

径级分布表 DBH Class

胸径区间 (Diameter class) (cm)	个体数 (No. of individuals in the plot)	比例 (Proportion)
1～2	25	0.33
2～5	21	0.28
5～10	22	0.29
10～20	7	0.09
20～30	0	0.00
30～60	0	0.00
≥60	0	0.00

152 茜树（山黄皮） qiànshù(shānhuángpí) | Maddertree

Aidia cochinchinensis Lour.
茜草科 | Rubiaceae

代码（SpCode）= AIDCOC
个体数（Individual number/24ha）= 9
最大胸径（Max DBH）= 3.8 cm
重要值排序（Importance value rank）= 128

常绿灌木或乔木，高2～15m。枝无毛。叶革质或纸质，侧脉5～10对。聚伞花序与叶对生或生于无叶的节上。浆果球形，熟时紫黑色。种子多数。花期3～6月，果期5月到翌年2月。

Evergreen shrub or tree, 2-15 m tall. Branchlets glabrous. Leaf blade leathery or papery, lateral viens 5-10 pairs. Inflorescences cymous opposite to leaves or at leafless nodes. Berry globose, dark purple at maturity, with many seeds. Fl. Mar.-Jun., fr. May to Feb. of next year.

植株 Whole plant
摄影：贾琪 Photo by: Jia Qi

小枝 Branchlet
摄影：贾琪 Photo by: Jia Qi

花序 Inflorescence
摄影：贾琪 Photo by: Jia Qi

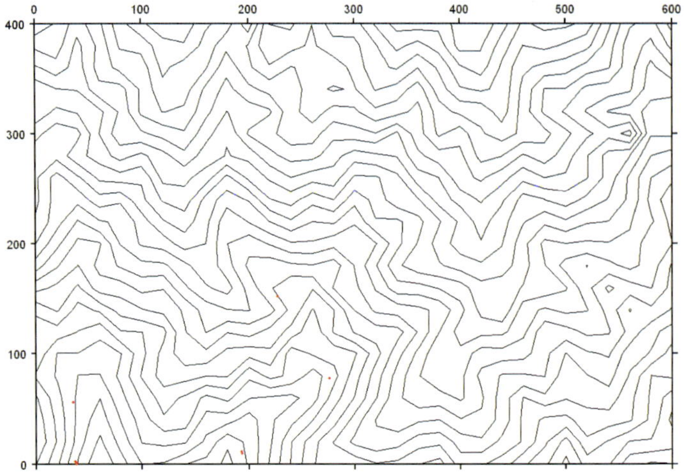

个体分布图 Distribution of Individuals

径级分布表 DBH Class

胸径区间 (Diameter class) (cm)	个体数 (No. of individuals in the plot)	比例 (Proportion)
1～2	5	0.56
2～5	4	0.44
5～10	0	0.00
10～20	0	0.00
20～30	0	0.00
30～60	0	0.00
≥60	0	0.00

153 鸡仔木 jīzǎimù | Chicken Wood

Sinoadina racemosa (Sieb. et Zucc.) Ridsdale
茜草科 | Rubiaceae

代码（SpCode）= SINRAC
个体数（Individual number/24ha）= 10
最大胸径（Max DBH）= 23.7 cm
重要值排序（Importance value rank）= 124

落叶乔木，高4～12m。树皮灰色，粗糙，小枝无毛。叶薄革质，侧脉6～12对。头状花序不计花冠直径4～7mm，常约10个排成聚伞状圆锥花序。果序直径11～15mm。花果期5～12月。

Deciduous tree, 4-12 m tall. Bark greyish, roughly. Brabchlets glabrous. Leaf blade thinly leathery, lateral viens 6-12 pairs. Inflorescence capitate, 4-7 mm in diameter without corona, always 10 make up cymous panicle. Infructescence 11-15 mm in diam. Fl. and fr. May-Dec..

枝叶 Branchlet and leaves
摄影：方腾 Photo by: Fang Teng

节 Node
摄影：方腾 Photo by: Fang Teng

花序 Inflorescence
摄影：王洪 Photo by: Wang Hong

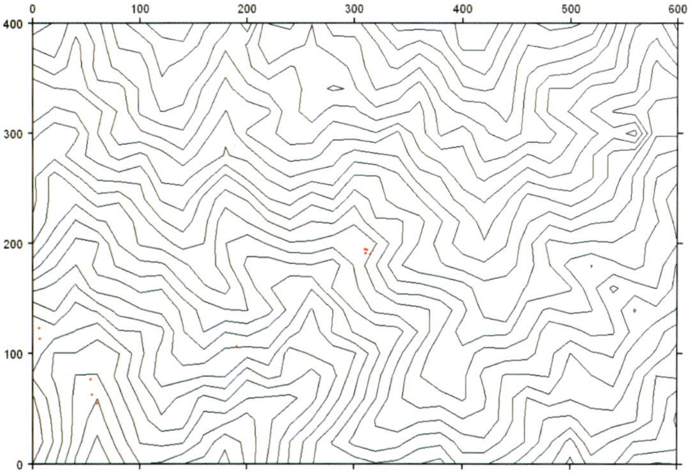

个体分布图 Distribution of Individuals

径级分布表 DBH Class

胸径区间 (Diameter class) (cm)	个体数 (No. of individuals in the plot)	比例 (Proportion)
1～2	3	0.30
2～5	4	0.40
5～10	0	0.00
10～20	1	0.10
20～30	2	0.20
30～60	0	0.00
≥60	0	0.00

154 白花苦灯笼 báihuākǔdēnglong | Whiteflower Tarenna

Tarenna mollissima (Hook. et Arn.) Rob.
茜草科 | Rubiaceae

代码（SpCode）= TARMOL
个体数（Individual number/24ha）= 146
最大胸径（Max DBH）= 3.8 cm
重要值排序（Importance value rank）= 69

落叶灌木或小乔木，高1～4m。小枝密被灰褐色柔毛。叶片薄纸质至纸质，上面密被短毡毛，下面密被柔毛。伞房状聚伞花序顶生。果实球形，直径5～6mm，被短柔毛。花期7～8月，果期9～11月。

Deciduous shrub or small tree, 1-4 m tall. Branchlets densely greyish brown pubescent. Leaf blade thinly papery to papery, adaxially densely short pubescent, abaxially densely pubescent. Inflorescence terminal, corymbose cyme. Fruits globose, 5-6 mm in diam, short pubescent. Fl. Jul.-Aug., fr. Sep.-Nov..

幼苗 Seedling
摄影：金冬梅 Photo by: Jin Dongmei

叶 Leaf
摄影：张金龙 Photo by: Zhang Jinlong

果枝 Fruiting branch
摄影：张金龙 Photo by: Zhang Jinlong

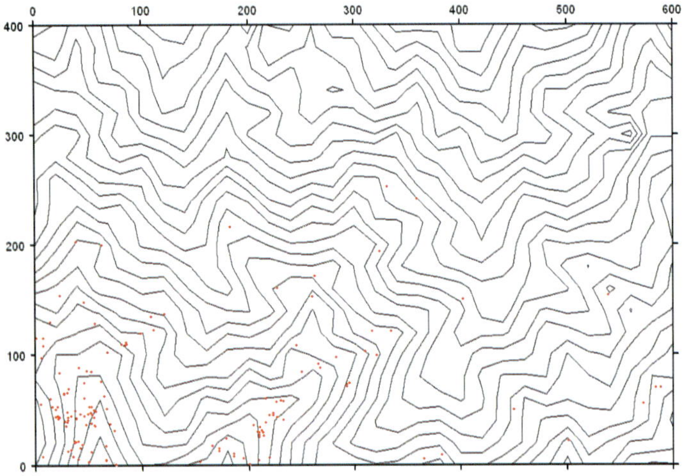

个体分布图 Distribution of Individuals

径级分布表 DBH Class

胸径区间 (Diameter class) (cm)	个体数 (No. of individuals in the plot)	比例 (Proportion)
1～2	143	0.98
2～5	3	0.02
5～10	0	0.00
10～20	0	0.00
20～30	0	0.00
30～60	0	0.00
≥60	0	0.00

155 狗骨柴 gǒugǔchái | Common Dogbonebavin

Diplospora dubia (Lindl.) Masam.
茜草科 | Rubiaceae

代码（SpCode）= DIPDUB
个体数（Individual number/24ha）= 93
最大胸径（Max DBH）= 4.3 cm
重要值排序（Importance value rank）= 74

常绿灌木或小乔木，高2～5m。一年生枝灰黄色，光滑，二年生以上枝树皮薄片状剥落，呈铁锈色。叶片近革质，两面无毛。花腋生。果近球形，成熟时橙红色。花期5～6月，果期7～10月。

Evergreen shrub or small tree, 2-5 m tall. First year twiglets greyish yellow, glabrous, older then 2 years turn rusty, scaly peel off. Leaf blade semi-leathery, glabrous on both surface. Inflorescence axillary. Fruit subglobose, orange when mature. Fl. May-Jun., fr. Jul.-Oct..

幼苗 Seedling
摄影：方腾 Photo by: Fang Teng

果枝 Fruiting branch
摄影：张金龙 Photo by: Zhang Jinlong

核果 Drupe
摄影：张金龙 Photo by: Zhang Jinlong

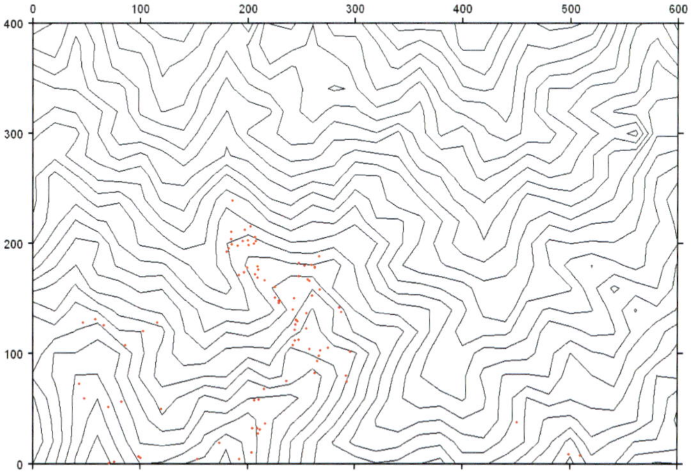

个体分布图 Distribution of Individuals

径级分布表 DBH Class

胸径区间 (Diameter class) (cm)	个体数 (No. of individuals in the plot)	比例 (Proportion)
1～2	58	0.62
2～5	35	0.38
5～10	0	0.00
10～20	0	0.00
20～30	0	0.00
30～60	0	0.00
≥60	0	0.00

156 宜昌荚蒾　　yíchāngjiámí | Yichang Arrowwood

Viburnum erosum Thunb.
忍冬科 | Caprifoliaceae

代码（SpCode）= VIBERO
个体数（Individual number/24ha）= 352
最大胸径（Max DBH）= 3.5 cm
重要值排序（Importance value rank）= 46

落叶灌木，高达3m。当年生小枝密被簇状短毛，二年生枝无毛。叶纸质，上面无毛，下面密被毛。复伞式聚伞花序生于具1对叶的侧生短枝之顶。果实红色，宽卵圆形。花期4～5月，果期8～10月。

Deciduous shrub, up to 3 m tall. First year twiglets densely tufted. Older branchlets trun glabrous. Leaf blade papery, adaxially glabrous, adaxially densely pubescent. Inflorescence compoundly cymous, terminal at 2-leafed lateral branchlet. Fruit red when mature, broad ovate. Fl. Apr.-May, fr. Aug.-Oct..

幼苗 Seedling
摄影：金冬梅 Photo by: Jin Dongmei

叶背 Leaf back
摄影：陈国科 Photo by: Chen Guoke

果序 Infructescence
摄影：陈国科 Photo by: Chen Guoke

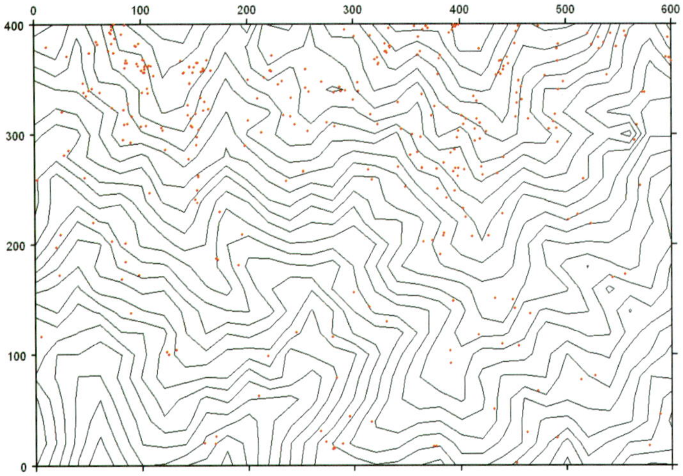

个体分布图 Distribution of Individuals

径级分布表 DBH Class

胸径区间 (Diameter class) (cm)	个体数 (No. of individuals in the plot)	比例 (Proportion)
1～2	261	0.74
2～5	91	0.26
5～10	0	0.00
10～20	0	0.00
20～30	0	0.00
30～60	0	0.00
≥60	0	0.00

157 具毛常绿荚蒾 jùmáochánglǜjiámí | Hariy Evergreen Arrowwood

Viburnum sempervirens K.Koch var. *trichophorum* Hand.-Mazz.
忍冬科 | Caprifoliaceae

代码（SpCode）= VIBSEM
个体数（Individual number/24ha）= 3
最大胸径（Max DBH）= 1.9 cm
重要值排序（Importance value rank）= 141

常绿灌木，高达4m。幼枝、叶柄和花序均密被簇状短毛。叶革质，侧脉5～6对，叶顶端具明显的锯齿。复伞形聚伞花序顶生。果实红色。花期5月，果熟期10～12月。

Evergreen shrub, up to 4 m tall. Young branchlets, petiole, inflorescence are densely pubescent. Leaf blade leathery, lateral viens 5-6 pairs, conspicuously serrate at apex. Inflorescence cymous compound, terminal. Fruits redish at maturity. Fl. May, fr. Oct-Dec.

叶 Leaf
摄影：张金龙 Photo by: Zhang Jinlong

枝 Branch
摄影：张金龙 Photo by: Zhang Jinlong

果序 Infructescence
摄影：方腾 Photo by: Fang Teng

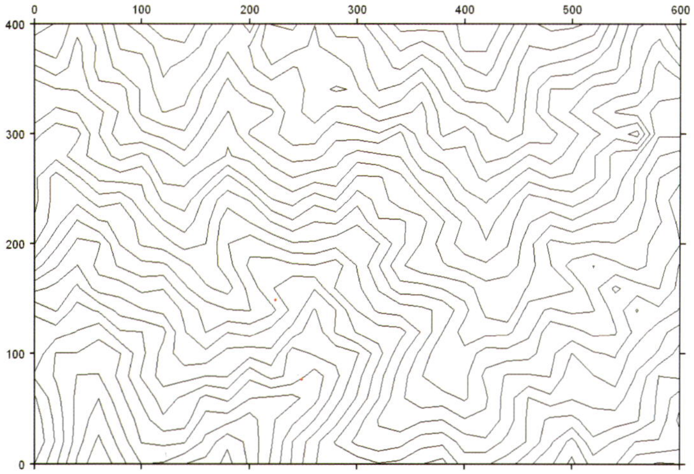

个体分布图 Distribution of Individuals

径级分布表 DBH Class

胸径区间 (Diameter class) (cm)	个体数 (No. of individuals in the plot)	比例 (Proportion)
1～2	3	1.00
2～5	0	0.00
5～10	0	0.00
10～20	0	0.00
20～30	0	0.00
30～60	0	0.00
≥60	0	0.00

158 茶荚蒾（饭汤子） chájiámí(fàntāngzǐ) | Tea Arrowwood

Viburnum setigerum Hance
忍冬科 | Caprifoliaceae

代码（SpCode）= VIBSET
个体数（Individual number/24ha）= 1
最大胸径（Max DBH）= 2.2 cm
重要值排序（Importance value rank）= 154

落叶灌木，高达4m。当年生小枝浅灰黄色，无毛。叶纸质，基部圆形。复伞形聚伞花序常弯垂。果序弯垂，果实红色，卵圆形，长8～10mm。花期4～5月，果熟期9～10月。

Deciduous shrub, up to 4 m tall. First year branchlets pale greyish yellow, glabrous. Leaf blade papery, base round. Inflorescence cymous compound, usually pendulous. Infructescence pendulous, fruits red at maturity, ovate, 8-10 mm long. Fl. Apr.-May, fr. Sep.-Oct..

叶 Leaf
摄影：郑小军 Photo by: Zheng Xiaojun

花序 Inflorescence
摄影：陈彬 Photo by: Chen Bin

果序 Infructescence
摄影：陈彬 Photo by: Chen Bin

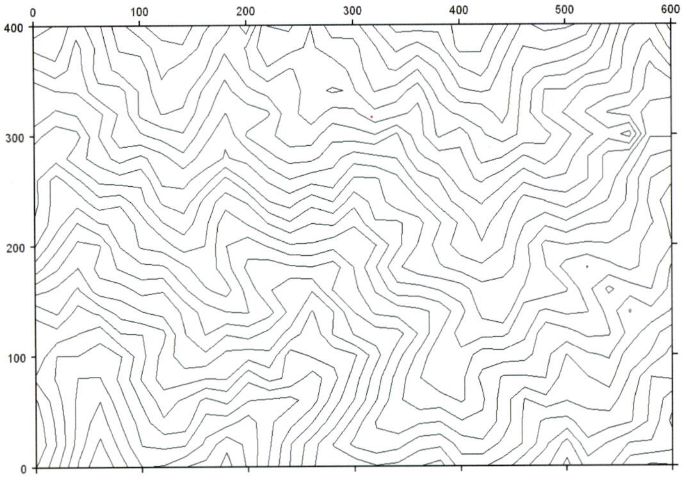

个体分布图 Distribution of Individuals

径级分布表 DBH Class

胸径区间 (Diameter class) (cm)	个体数 (No. of individuals in the plot)	比例 (Proportion)
1～2	0	0.00
2～5	1	1.00
5～10	0	0.00
10～20	0	0.00
20～30	0	0.00
30～60	0	0.00
≥60	0	0.00

159 水马桑

shuǐmǎsāng | China Weigela

Weigela japonica Thunb. var. *sinica* (Rehd.) Bailey
忍冬科 | Caprifoliaceae

代码（SpCode）= WEIJAP
个体数（Individual number/24ha）= 1
最大胸径（Max DBH）= 1.5 cm
重要值排序（Importance value rank）= 156

落叶灌木，高达6m。幼枝四方形，有2列柔毛，老枝无毛。叶纸质，下面密生短柔毛。聚伞花序1～3花，生于短枝叶腋或顶端。蒴果狭长。花期4～5月，果期8～9月。

Deciduous shrub, up to 6 m tall. Young branchlets square, with 2 rows of pubescence; old branch glabrous. Leaf blade papery, abaxially densely and shortly pubescence. Inflorescence 1-3-flowered, axillary or terminal at short branchlets. Capsule narrow. Fl. Apr.-May, fr. Aug.-Sep..

叶 Leaf
摄影：金冬梅 Photo by: Jin Dongmei

花 Flower
摄影：金冬梅 Photo by: Jin Dongmei

蒴果 Capsule
摄影：陈国科 Photo by: Chen Guoke

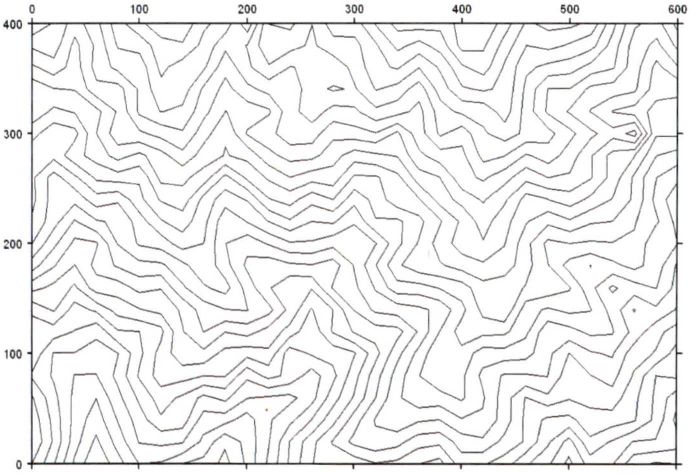

个体分布图 Distribution of Individuals

径级分布表 DBH Class

胸径区间 (Diameter class) (cm)	个体数 (No. of individuals in the plot)	比例 (Proportion)
1～2	1	1.00
2～5	0	0.00
5～10	0	0.00
10～20	0	0.00
20～30	0	0.00
30～60	0	0.00
≥60	0	0.00

附录I、植物中文名索引
Appendix I. Chinese Species Name Index

附录II、植物学名索引
Appendix II. Scientific Species Name Index